A Critical Analysis of Climate Change Opinion News in Malaysian Media

Nor Fariza Mohd Nor · Tasha Erina Taufek ·
Jamaluddin Aziz

A Critical Analysis of Climate Change Opinion News in Malaysian Media

palgrave
macmillan

Nor Fariza Mohd Nor
Center for Research in Language
and Linguistics, Faculty of Social Sciences
and Humanities
National University of Malaysia
Bangi, Selangor, Malaysia

Tasha Erina Taufek
Center for Research in Language
and Linguistics, Faculty of Social Sciences
and Humanities
National University of Malaysia
Bangi, Selangor, Malaysia

Jamaluddin Aziz
Center for Research in Media
and Communication, Faculty of Social
Sciences and Humanities
National University of Malaysia
Bangi, Selangor, Malaysia

ISBN 978-3-032-10332-1 ISBN 978-3-032-10333-8 (eBook)
https://doi.org/10.1007/978-3-032-10333-8

Cover credit: Alistair Berg/GettyImages

This Palgrave Macmillan imprint is published by the registered company Springer Nature Switzerland AG
The registered company address is: Gewerbestrasse 11, 6330 Cham, Switzerland

Preface

This may sound cliché, whether we like it or not, we are all living with climate change or climate crisis. The experiences, which used to be part of our mediatised experience—it is always elsewhere!—are rapidly becoming our daily lived experiences and realities. The Anthropocene is no longer a far distance concept; it is a bitter reality that we need to face head on and with all honesty. We realise that the history of the present moment shapes our fundamental inquiries: How do we arrive at this present state of climate change? What helps shape our perception and attitude towards climate change locally and globally?

As our knowledge as a team intersects with our expertise in discourse analysis, we often wonder how language (as a tool and a symbol) in the media is used to effect change in the perceptions and attitudes of Malaysians. We wanted to make sense of how, in the context of mis/information influx and post-truth, the power of language and the behaviour of the media work to bring climate change to our consciousness as a nation. Importantly, we seek proof of how ideology keeps structuring our perception through language and reproducing injustices via the media system.

Our understanding of climate change as a historical present also reflects our awareness of the power of the media as an ideological apparatus—to use Louis Althusser's salient concept—that can be influential in forming our belief and norms regarding climate change. Simply put, we ask: How does the hegemonic power of the (mainstream) media work to create common sense about climate change? How different or similar are the roles of the alternative/tabloid media from that of the mainstream media? We pursue this line of inquiry with a caveat that we may "poke at the hornet nest" of the policymakers.

If we could paint the picture of our research journey and the writing of this book, we would paint a picture of undulating hills. There are ups and downs; undeniably, they are necessary for us to grow as scholars and to be "uncomfortable" in our pursuit of knowledge. Over hundreds of cups of coffee and tea, we argue, draft and redraft our work, deal with technical glitches, and strategise/re-strategise to accommodate contexts and changes in our journey. This we do amidst chasing a miasma of academic key performance indicators that we have to fulfil every year.

We are grateful because our proposal entitled Dynamic Word Embeddings Ecolinguistics model (DWEEM) on Public Perceptions and Awareness about Climate Change submitted to the Ministry of Higher Education, Malaysia, in 2020 was accepted and the Fundamental Research Grant (FRGS) we received modulates fundamental inquiries into a larger research scale. The grant makes this research possible at this scale.

This book is meant for readers who are interested in discourse analysis, Critical Discourse Analysis, Corpus Linguistics, ecolinguistics, environmental studies, and Media Studies. General readers may find this book useful as we do try to have them in our mind when writing this book. We particularly hope this book is useful for students and scholars of language, linguistics, and media studies.

Bangi, Selangor, Malaysia

Nor Fariza Mohd Nor
Tasha Erina Taufek
Jamaluddin Aziz

Acknowledgement We wish to extend our sincere appreciation to Universiti Kebangsaan Malaysia (UKM), our home institution, for providing the intellectual environment and institutional support that enabled the completion of this work. We also gratefully acknowledge the Ministry of Higher Education (MOHE), Malaysia, for funding this project under the Fundamental Research Grant Scheme (FRGS/1/2020/SS10/UKM/02/1). We are also indebted to the interview participants whose expertise and perspectives greatly enriched our study, in particular representatives from Amanah Lestari (ALAM), Bank Pembangunan Malaysia Berhad (Malaysian Development Bank Limited), the Water Resources Division, Ministry of Natural Resources, Environment and Climate Change, and the Institute of Climate Change, Universiti Kebangsaan Malaysia (UKM). Their generous contributions of time and knowledge were invaluable in shaping the analyses and insights presented in this book.

Competing Interests The authors have no competing interests to declare that are relevant to the content of this manuscript.

Contents

About the Authors

Nor Fariza Mohd Nor is a Professor at the Center of Research in Language and Linguistics, Faculty of Social Sciences and Humanities, Universiti Kebangsaan Malaysia (UKM). She has led numerous research grants from the Ministry of Education, industry and international organisations. Her research primarily focuses on discourse studies, with a particular focus on media discourse and corpus-assisted discourse studies. Her work spans areas such as the Malaysian Hansard corpus, Malay culturomics, the Malaysian climate change corpus, and social media corpus. She has published extensively in high impact and indexed journals, published book chapters and co-authored several books, including **Action** Research: A Window to New Experiences in Teaching and Learning (2017), Undergraduate Research in The Social Sciences: Practices and Perceptions (2017) and Understanding Culturomics (2023). She also serves as the editor for Heliyon, a high-impact Quartile 2 journal, and as a peer reviewer for other high-impact journals.

Tasha Erina Taufek is a Ph.D. candidate and emerging scholar in English Language Studies at the Centre for Research in Language and Linguistics, Faculty of Social Sciences and Humanities, Universiti Kebangsaan Malaysia (UKM). Supported by the Ministry of Higher Education through the prestigious Fundamental Research Grant Scheme (FRGS), her doctoral research breaks new ground with a corpus-assisted discourse study of climate change issues in Malaysia. Her expertise spans corpus linguistics, discourse analysis, ecolinguistics, and sentiment analysis, with a particular emphasis on the interplay between media, language, and public engagement with the climate crisis. A rising voice in the field, she has presented her work at academic conferences and contributed to scholarly publications, advancing critical conversations on how discourse shapes environmental awareness, policymaking, and collective action.

Jamaluddin Aziz is an Associate Professor in Critical Media Studies, at the Centre for Research in Media and Communication, the National University of Malaysia. His research, publication, and advocacy interests include media, cultural and critical theories, gender analysis, media discourse, leadership communication, digital and environmental humanities, visual culture, and film studies. His latest edited book is entitled Understanding Culturomics (2023). He has also published widely in journals such as Feminist Media Studies and International Journal of Communications (IJOC). He was appointed as the Gender Expert for the Ministry of Women, Family and Community Development (2015-2016). He worked with MARA (The Council of Trust of the People) from 2022 to 2025 to develop and launch MARA Communities Women Empowerment Action Plan and was also involved in Men Health Policy (2020-2025) with Malaysian Ministry of Health. He now sits with the MARA women empowerment committee. He is now the chief editor of Akademika: Southeast Asian Journal of Social Sciences and Humanities, a section editor of GEMA online. At present, he is the President of the Association of the National University of Malaysia Academics.

List of Tables

1

A Multidimensional Climate Change Discourse: Media, Stakeholder Perceptions, and National Context

Introduction

This chapter introduces the readers to the context of this book by locating it within the global and local milieux, while foregrounding the significance of understanding environmental issues via the media. It begins by underscoring the media's pivotal function in influencing audience attitudes and behaviours through the interpretation and dissemination of climate change news. The discussion presents the nuanced power of the media and its intricate role, recognising its potential to mould public understanding and contribute to polarisation on the issue. Despite the complexity involved, the chapter underscores the significance of the media as a critical source of information about climate change throughout the globe. Transitioning from the media's influential role, the chapter progresses to an examination of the climate change landscape in Malaysia. It provides an overview of research on climate change in Malaysia, shedding light on the vulnerability of the country to climate impacts. The exploration into media influence and Malaysia's climate challenges enriches the reader's understanding of the multifaceted dynamics surrounding climate change discourse. This chapter ends with

N. F. M. Nor et al., *A Critical Analysis of Climate Change Opinion News in Malaysian Media*, https://doi.org/10.1007/978-3-032-10333-8_1

the analysis of the interview data gathered from stakeholders in Malaysia, such as fellow academicians, private and government sectors.

Climate change is an issue filled with contradiction and this irony is undergirded by climate change discourse. For starters, the concept of climate change is constantly debated and challenged. For some experts, the term climate change fails to capture the urgency and currency of climate disaster or catastrophe, arguing that the term climate change reinforces the idea that it is part of an evolution—hence natural and necessary—rather than the result of human intervention. On the other hand, the term climate change is also framed as "urgent and long term" "personal and institutional", and "local and global" all at once. Needless to say, this dilutes the understanding of climate change itself.

Governments, driven by their own socio-economic motivations, often use the problem of definitions to continue their agenda of national sovereignty. Karkour (2023, p. 317) argues that nationalism has "twofold negative relationship, where nationalism exacerbates climate change and climate change exacerbates nationalism". This means that the issue of climate change can be framed, and therefore, manipulated by the powers that be, which usually is the elitist. The Indonesian President, Prabowo Subianto, recently claimed that the expansion of oil-palm plantations poses no threat to forests because oil palm is a tree and does not lead to deforestation. Prabowo's remarks received severe criticisms. Critics and scientists argue and warn that peatlands in oil palm plantations which store massive amounts of carbon, are often drained and burned for oil palm plantations, hence, releasing substantial amounts of greenhouse gases. Environmentalists caution that such rhetoric could legitimise further deforestation and encourage security-enforced land grabs. Internationally, these statements risk undermining Indonesia's credibility in climate negotiations, and could jeopardise access to green investments and key export markets like the EU.

As nationalism relates to the question of sovereignty and national borders, communications about the need for infrastructure and new township pivot around rambunctious globalisation and Western hegemony. The discourse of national and the rights for modernisation—often played out in the government-owned media organisations—obfuscating the dire need to protect the environment. Likewise, the argument that centres on "government is doing anything it can" speaks about how climate change affects nationalism and governance. Nationalism then is often used to challenge the global pressure and discourse on climate change; thus, blurring the fact that national boundaries are irrelevant in climate change.

Globally, climate communication especially via legacy media epitomises global news biasness; resulting in more decolonising narratives, which resist Western dominance, proscribing Western hegemony. In the West, climate communication often involves 'greenwashing' or 'whitewashing', where companies and governments speak about sustainability while ignoring harmful practices. Terms such as "net-zero" and "climate neutrality" are often used rhetorically to enhance reputations without measurable results. In most cases, the scientific discourse of "net-zero" and "climate neutrality" is often shrouded in confusing and misleading jargon, deliberately designed to create moral uncertainty. Meanwhile, others, particularly in the Global South (if the term is still relevant), are branded as environmental criminals who ignore colonial and capitalist histories of extraction. This dual discourse, where climate change is branded for some and weaponised for others, reflects ongoing power imbalances around who can speak, act, and be heard.

Given that in the trajectory of the Anthropocene—an epoch in which human activities have determined environmental and geological processes—it is now understood that climate information is not always neutral or objective but rather filtered through cultural, political, and institutional lenses. Traditionally, the media landscape is the embodiment of such triadic perspectives, which are crucial in shaping people's perception of climate change issues. Varied types of discourses are often applied, depending on the agenda set by the institution. As an example, science and technology discourses can be framed differently for different audiences/agendas. Another instance is when media reporting in some regions might stress economic threats more than environmental justice, or technological optimism more than systemic critique often to appeal to specific groups of media audiences. This discursive bias shapes public opinion and can drive action or cause apathy. In addition, access to climate data and scientific literacy varies widely, which creates uneven engagement across populations, and this intensified media ecology.

Conterminous with contemporary media ecology, scholars have generally agreed that there are three shifts of communication paradigms, informing the way media and communications are intellectualised. The first paradigm is known as the effect paradigm. The main assumption in this paradigm is that the media has a powerful effect on the audience. This paradigm comes into being through the study of media propaganda after World War I. Then it shifts to the second paradigm, which reflects nations coming to terms with post-war disaster, and focusing on nation building—hence the term development communication. Within this paradigm, the media is used to inform the general public of the government's initiatives, and to communicate about the idea of re/building the country. The third paradigm is the critical paradigm, which utilises a critical approach to understanding media landscape and behaviour. Analysing the media using Critical Discourse Analysis is an example of how media is examined critically, and the purpose is to excavate the media texts' ideological imperative, which is the general aim of this book.

We presented in the early section of this chapter a brief review on the state of global media discourse and government stances on climate change before transitioning to a specific discussion on past studies on climate change from the global context of media, to past research on climate change in the Malaysian media. Interview excerpts from the stakeholders follow next to highlight the government's plan and policies on the state of climate change in Malaysia. In so doing, the chapter alludes to historical moments that helped create these concerns and brings climate change to global consciousness. Sparks (2007, p. 8) argues that "[n]o ideas ever do evolve without reference to the times in which they are developed". By linking climate change issues with the media, the book inevitably provides a critical analysis of discourse on climate change. It does this by revealing how the media and its power can be manipulated in the process of production, dissemination, and consumption of news.

This book takes the position of advocating the discourse change while keeping in mind that the terms may have diverse meanings and implications. The term climate change and crisis will be used interchangeably. We argue that climate change discourse in the Malaysian news media plays crucial roles in disseminating and creating consciousness about the environment. For that reason, we focus on Malaysian opinion news as our corpus, which will be supported with examination of linguistic strategies using corpus linguistics approach, attitude analysis using the Appraisal Theory (Martin and White, 2005), critical discourse analysis approach (Fairclough, 1995) and Stibbe's ecolinguistics framework (2015).

Global Media and Climate Change

The way in which climate change is communicated is of immense importance to raise public awareness and promote actions. Research in media and communication studies have made immense contributions since the 1990s in revealing how climate change has been communicated to the mass. Therefore, the media's role in shaping public awareness and understanding of climate change is pivotal, extending beyond mere information dissemination. It acts as a medium that translates scientific data that may be complex to understand by the public into laymen's narratives that can be better understood, hence influencing public perception and action. Carvalho (2010) asserts that media are important agents in producing, reproducing, and transforming the meanings of particular social issues.

The transformative power of media in shaping how society perceives and responds to the climate crisis is undeniable. Climate change discourse, as Newman et al. (2020) highlight, is a complex and unobtrusive issue, often obscured by its long-term nature and intricate scientific underpinnings. The media, therefore, plays the role of making these scientific complexities accessible to the public, for instance, by highlighting the urgency and relevance of climate action. Furthermore, Sampei and Aoyagi-Usui (2009) demonstrate that increased news media attention directly correlates with heightened public concern. This includes framing climate change not just as an environmental problem but also as a social, economic, and political one, thus broadening its appeal and impact. Therefore, consistent and robust reporting is crucial for maintaining public engagement and driving action. Effective media coverage involves not only reporting scientific findings, but also showcasing the human vulnerabilities behind climate change, highlighting the impacts on vulnerable communities, thus, raising public awareness and encouraging the public to be involved in helping these vulnerable communities. For instance, investigative journalism exposing corporate greenwashing (a term coined by environmentalist Jay Westerveld in the 1980s), or reporting on successful renewable energy projects can significantly influence public opinion and policy. Moreover, the media's role that extends to scrutinising the actions of governments and corporations, holding them accountable for their climate commitments can foster an understanding of climate change that integrates diverse voices, including scientists, activists, and affected communities. Consequently, the media's role in reporting on climate change can empower individuals and communities to take meaningful action.

The media's portrayal of climate change is significantly shaped by the interplay of domestic and global perspectives, influencing how causes, impacts, and solutions are framed. Climate change coverage is not only a media report, but rather a reflection of prevailing societal understandings (Konieczna et al., 2014; Kunelius et al., 2017; Olausson, 2009). This means coverage on climate change is dependent on the domestic and global perspectives on what causes climate change, who is affected, or how it can be addressed. These understandings are intertwined with political and economic contexts, leading to diverse media narratives across different nations. Political and economic factors, such as national climate policies, economic vulnerabilities, and ongoing climate actions, play a crucial role in shaping media coverage (Barkemeyer et al., 2017; Schmidt et al., 2013; Vu et al., 2019). Past studies have revealed differences in climate change communication across countries, political contexts, media platforms, and time periods. For instance, countries with established climate action plans tend to have media reports that emphasise solutions and urgency, while nations prone to climate change impacts (e.g., small island states, developing countries) often have media reports that focus on the immediate and devastating consequences. From another perspective, countries that are heavily reliant on fossil fuels may exhibit media coverage that restraints the urgency of climate action or emphasises economic trade-offs. Conversely, nations experiencing severe climate impacts may prioritise reporting on the immediate consequences and promote stronger mitigation measures.

Carmichael et al. (2017), Kim (2011), and Newman et al. (2018) studies revealed that news in the United States used partisan framing, whereby climate change is presented through political lens that reinforce pre-existing beliefs and foster polarisation. Chinn et al.'s study (2020) showed a significant trend in U.S. climate change news from 1985 to 2017, where political discourse increasingly overshadows scientific discourse. This shift signifies a significant change in how climate change is framed for the public. They found that Republican mentions outnumbered Democratic mentions, intensifying polarisation in climate change coverage. This imbalance suggests that political narratives of the Republican party have disproportionately shaped public perception, potentially leading to a skewed understanding of the scientific consensus. The politicisation often results in the spread of misinformation, further complicating public understanding of climate change. Weiner et al. (2021) examined the linguistic markers of climate change news in six U.S. newspapers, focusing on proximity cues—markers of temporal and spatial distance. Their findings indicate that while proximity cues were infrequent, proximal cues (indicating nearness) were more prevalent than distal cues (indicating distance). The tendency towards highlighting immediate, local impacts rather than distant, future consequences can influence how readers perceive the urgency and relevance of climate change. The emphasis on proximal events, such as local extreme weather in local news may foster a greater sense of immediacy and personal connection, potentially motivating local action. However, the infrequent use of any proximity cue indicates that many news outlets still struggle to effectively convey the scale and scope of the climate crisis. The lack of distal cues may also lead to a lack of understanding of the global and long-term effects of climate change among the public.

Likewise, Happer and Philo (2013, 2016) conducted a study on news reports in the United Kingdom. They described how newspapers, television, and broadcast news in the United Kingdom selectively highlight specific aspects of climate change while marginalising alternative viewpoints. This deliberate approach shapes public perception, effectively controlling the narrative and limiting exposure to diverse perspectives and agendas that focus public attention on particular facets of climate change while often undermining alternative perspectives. The deliberate decision to prioritise certain narratives, may inadvertently reinforce existing biases, where dissenting opinions are silenced. Additionally, this targeted agenda-setting can influence public discourse by focusing on often sensationalised climate change aspects, hence, contributing to an only informed public, but perhaps not an educated one. Eskjær (2013) documented regional differences—with studies from the Middle East, Scandinavia, and North America in which media use scientific uncertainty, urgency, and relevance framing, that can create public confusion and disengagement because climate change is portrayed as an immediate threat. However, Thaker et al. (2017) noted that consistent, accurate coverage in India can boost public engagement. Results suggest that the Indian media, through consistent and accurate coverage of global warming using trusted sources, managed to play a positive role in increasing public engagement among a largely unaware population in India.

The media, therefore, acts as a crucial platform for the construction of climate change narratives. Past studies have highlighted how the media chooses to highlight certain aspects of climate change, and downplay others, such as choosing to focus on extreme weather events, or on the progress of renewable energy technologies. This can have a profound impact on public perception. The media's ability to frame climate change as an immediate threat or a distant problem significantly influences public engagement and policy support. Furthermore, the media's role in disseminating scientific information and fostering public discourse is essential for informed decision-making.

Malaysian Media and Climate Change News

The media in Malaysia plays an important role in shaping the public's opinion regarding climate change. Arguably, Malaysian media is also influential in influencing public policy, while raising Malaysians' consciousness about climate change issues. Past studies on climate change in the Malaysian media used printed English newspapers, Malay newspapers, Chinese language newspapers, and two online newspapers as data. Azmi et al. (2015) used a content analysis approach to examine how two mainstream Malay language newspapers and two mainstream English language newspapers presented reports on climate change. The results revealed that the trend of the newspapers' coverage on global climate change in Malaysia mainstream newspapers correlate with the major events related to the issue. This means that newspaper coverage on any global climate change issue increased drastically during major events and would consequently decrease when no major climate change events are reported. Manzo and Padfield (2016) conducted a content analysis of climate change reports in five English language newspapers over a three-year period (2009–2011) to find out how climate change is framed in the Malaysian media. The results show that climate change has been framed as both a multi-scalar responsibility and a positive opportunity for two key stakeholders in development, i.e. neoliberal market forces and geopolitical actors who are interested in restructuring the international political economy. Additionally, two key themes emerged which are climate capitalism and green nationalism. The two discourses actually interact frequently with emerging economic discourses. Cheong and Zanuddin (2018), and Haris et al. (2022) found a focus on policy-driven coverage, with attention to mitigation, adaptation, and environmental non-governmental organisations' activities in two mainstream Chinese language newspapers. Manzo and Padfield's (2016) study on the coverage of climate change in two Chinese newspapers revealed that climate change news is affected by global coverage of major climate change events. Harris et al. (2022) conducted a systematic literature review of 19 published articles from Scopus and Web of Science-indexed to understand the current state of the Southeast Asian Environmental NGOs participation in climate change governance. The analysis indicate that the Environmental Non-Governmental Organizations (ENGOs) in Southeast Asia are involved both directly and indirectly in climate change governance. In fact, the ENGOs play a significant role in the implementation of the climate change policy. However, they only have a minimal role in the formulation of said policy. Haris et al. (2022) conclude that the ENGOs could be an indispensable partner to the government in the climate change governance process as they can bring effective policy improvements.

The reviewed past studies on climate change in the Malaysian media demonstrate that in the Malaysian context, the medium is the message, to use Marshall McLuhan's famous expression. Mohamed et al. (2020) examined how mainstream newspapers in Malaysia (two Malay language newspapers and two English language newspapers) frame sustainable energy as an indication of how the issue is perceived by the public. They conclude that English newspapers and Malay newspapers have slightly different focus. English newspapers have more coverage on sustainable energy with sources, news types, and frames that indicate the importance and urgency of the issue, while Malay newspapers depend expansively on governmental agencies and journalist reporting, hence limiting the context of the issue. Rather, Malay newspapers—driven by political and economic imperatives—often frame climate-related issues by alluding to multifarious working-class apprehensions such as unemployment and food security. This denotes differences in framing news about sustainable energy between English and Malay newspapers. Indeed, it can be said that a more "Western" educated population—the one who typically reads English newspapers—is more interested in climate change issues and is more critical of the government vis-à-vis the Malay newspaper readers. This can be explained by the idea that some of the news are sourced from international news agencies, like Reuters and BBC, which are typically more critical of climate issues.

Meanwhile, Taufek et al., (2022) employed a thematic analysis approach to identify themes in news on climate change and the extent of media propagation about climate change. Their study concludes that the media reports trend on climate change is inconsistent because the coverage about climate change is dependent on the domestic and global perspectives on what causes climate change, who is affected, or how it can be addressed. Three themes on climate change are identified, which are governance, mitigation and adaptation, and threats. The study concludes that the media plays a very pertinent role in disseminating news on climate change to enhance public engagement and behaviour towards climate change. However, the study does not indicate how effective such reporting is in bringing climate change awareness to Malaysians.

What must be a point of contention with regards to the effectiveness of the media raising public awareness of climate change is the very idea of what kind of awareness is being perpetuated. This is due to the fact that in the Malaysian context, news media is "owned" or heavily regulated by the government, making it clear that awareness is coloured by political-economic imperatives. For instance, Hassan et al. (2023) study reported that disinformation about climate change is more likely to be spread by politicians, organisations, and anonymous agencies. Politicians appear to fabricate contents about climate change, hence, spreading disinformation which influences the public's understanding and perceptions about climate change. The study calls for joint efforts of activists, media practitioners, and governments to mitigate disinformation about climate change.

Based on these past studies, the patterns of climate change coverage in the Malaysian media can be categorised into three; (1) environmental vs development narratives (Manzo & Padfield (2016), (2) political and economic framing (Hassan et al., 2023; Taufek et al., 2022), and (3) local vs. global context (Azmi et al., 2015; Haris, 2025). Furthermore, there is significant variation in the types of climate change report, coverage, and framing across English, Malay, and Chinese newspapers, with English language sources generally being more proactive.

Stakeholder Perceptions

While this book analyses opinion news in Malaysian news organisations, there is a need to understand Malaysian stakeholders' perspectives on the issue of climate change. We interviewed three respondents who are key figures in their respective sectors. The interviews were carried out separately due to the availability of the respondents. For clarity purpose and for maintaining the nuances of each response by the respondents, the interview is kept verbatim. Editing is made only for clarity and grammatical accuracy. The respondents are a researcher and a senior lecturer at the Institute of Climate Change at a public university in Malaysia, a senior officer at the Water Resources Division, Ministry of Natural Resources, Environment and Climate Change Malaysia, and a corporate

Table 1.1 Respondents information

Name	Affiliation
Ms. Fatimah	Amanah Lestari (ALAM), Bank Pembangunan Malaysia Berhad (Malaysian Development Bank Limited)
Madam Siti	Water Resources Division, Ministry of Natural Resources, Environment and Climate Change
Dr. Chee	Institute of Climate Change, The National University of Malaysia (UKM)

figure who is the CEO at the Climate Change and Sustainability Department, at a bank in Malaysia. The respondents' actual names are withheld and pseudonyms are used instead. Table 1.1 provides a brief description of each respondent.

The inclusion of interview excerpts from the three key Malaysian stakeholders—a climate change researcher and expert, a policymaker, and a CEO provides a contextual lens for the book's critical analysis of climate change discourse in opinion news. The interviews are vital to demonstrate that climate change is not a theoretical or distant issue in Malaysia, but a reality shaped by the diverse and often competing perspectives of those at the forefront.

The interviews help the reader understand the research by bridging the gap between media discourse and real-world action. While opinion news offer a glimpse into public discussions and views on climate change, the expert interviews provide formal and systemic insights into the challenges, opportunities, and political will surrounding climate action. In addition, these interviews offer an insight into the state of climate mitigation initiatives in Malaysia. The researcher provides a scientific and academic viewpoint, the policymaker offers insights into governance and implementation, and the CEO reveals the practicalities and motivations of the industries.

The CEO from the Malaysian Development Bank Limited was chosen as one of the respondents because the bank has a department named Amanah Lestari ALAM (ALAM) that specifically aims to transform Malaysians' mindset and behaviours towards environmental sustainability. ALAM (which means nature) aims to create awareness, educate, and instil a whistleblowing culture by collaborating with the public and private sectors, as well as communities on climate change issues and activities. ALAM is a national initiative mooted by the Ministry of Finance Malaysia and spearheaded by Bank Pembangunan Malaysia Berhad (Agro Bank).

Meanwhile, the Water Resources Division, Ministry of Natural Resources, Environment and Climate Change was selected as one of the stakeholders because the respondent can provide insights into the government's plans and policies (present and future) about climate change. This is vital in understanding the government's initiatives and efforts in Malaysia and at the global level on climate change.

In addition, the Institute of Climate Change at the University of Malaysia (UKM) was selected as one the stakeholders because the respondent is actively involved in research and can provide insights into climate change situations in Malaysia and abroad.

Current climate change trends and critical issues were discussed and two themes emerged from the interview, which are policy implementation challenges, and tension between mitigation and adaptation efforts. This section is organised along these themes. The responses for each question by the respondents are followed by the authors' comments and remarks. These comments and remarks are useful in summarising the responses together.

Theme 1: Policy Implementation Challenges

Policy implementation challenges are challenges related to the implementation of the policy. The most dominant theme obtained from the interview is the challenges in policy implementation. The implementation of climate change policies plays a critical role not only in driving local environmental action, but also in positioning the country as a responsible global actor—especially because it is signatory to the Paris Climate Agreement. When effectively enforced, these policies can elevate Malaysia's reputation on the global stage and foster international collaboration and investment in sustainable development. Nonetheless, implementing climate change policies is not without its challenges.

Researcher: Can you share with us a little bit about your research and experience? You informed me that you were from the Climate Change

Division (CCD) previously, right? Perhaps you can share with us the function of that division and a bit about yourself?

Dr Chee: My research is mainly on weather and air quality prediction. After joining the institute of climate change, I have to do more on the climate change part—how does climate change impact on the weather and air quality in general?

Researcher: Do you notice any difference in terms of climate change, weather, and air quality situations from five years ago and now? Does the Malaysian government consider the climate change guidelines and policy globally? Has there been any change in the policy on climate change in Malaysia, for instance, in the past ten years?

Dr Chee: We do see the temperature has increased by around 1.1-degree celsius over the past 20 years. It is data collected from Malaysia [sic]. We do see that trend and we do see the ozone [sic] is increasing also. When talking about climate change, we need at least 20 years of dataset. Since we are a tropical wet climate zone, basically we are experiencing wet and dry conditions or hot or not that hot [sic]. We are feeling more extremities—a lot of rainfall or dryness. Our raindrops are different from other countries, we reflect a lack of water supply or fire burning[sic] that creates a lot of emissions from air pollution. Extreme rainfall will become flooding [sic]. That is the problem that Malaysia is currently facing.

Madam Siti: Basically, all the climate actions are driven by the parties themselves. For instance, in the Paris Agreement, there are parties who would ratify them and then you have to fulfil the obligations (e.g., reducing emissions) and also to take actions to mitigate the impacts of climate change… You must know that when we talk about multilateral agreement, there are multilateral obligations. We negotiate as a block of countries. At national level [sic], clearly, that is what the division has come up with—we're waiting for the long-term low emission development strategy (LT-LEDS). For the local level, it's basically embedded in our green development initiatives—that's the green masterplan and all that [sic]. The local level initiative is not translated directly into the sectors for inventory.

Paris is the second Multilateral Environmental Agreement (MEA) for climate change. The first one would be the Kyoto Protocol. The US actually decided to pull out from Kyoto and then we have the Paris Accord.

Then, they negotiated for the Paris Agreement in 2015 and if I'm not mistaken, Malaysia was the first Southeast Asian country to ratify it. Part of Paris requirements is the obligation to reduce emissions (article 2) and also to build resilience against the impacts. For that, we need to report [sic]. Basically, when you reduce emissions, the baseline would be the inventory of the greenhouse gas emissions. We know basically what we need to do, but a lot of things are taking a bit of time to marginalise.

Ms. Fatimah: The trend for addressing climate change in Malaysia has rapidly evolved compared to previous years. Now, various policies related to sustainability have emerged. Malaysia ratified the United Nations Framework Convention on Climate Change (UNFCCC) in 1994 and subsequently ratified the Kyoto Protocol in 2002. As a climate action step, Malaysia formulated the National Climate Change Policy in 2009. In the same year, Malaysia also established the National Renewable Energy Policy. Since then, Malaysia has engaged in various acts and policies to address climate change issues. Malaysia signed the Paris Agreement to combat global warming in 2015. Global technology evolves over time, and we cannot be left behind, hence, in 2017, the Malaysia Green Technology Agenda 2030 was introduced.

A Specific Question to Madam Siti About the Public Awareness and Reaction to Climate Change

At the moment, public knowledge about climate change is flooding. The whole cycle of cause and effect is not really understood by them. You also mentioned that the market thinking is different from the policy. How can the general public change this thinking?

Madam Siti: The reason why the public equates climate change to flooding in Malaysia is because the impacts are felt and it is tangible. Rather than you talk to the public about transitioning to green farming etc. [sic], it's not relatable to them. In fact, there is a Merdeka survey pertaining to the environment that found that the Malaysian public can only relate to climate change when there are significant events such as haze, etc. If we were to tell them that we want to reduce electricity subsidies to help climate change, the public will be angry. It's different from the Western approach. Also, for example, if we try to change the term 'global warming' to 'global boiling', maybe UKM (our institution) can help us to create a better narrative of it for the public. Basically, if the public does not have enough to survive, they can't relate to thinking about climate change.

Madam Siti: We have a climate change policy. Unfortunately, it was only 38 pages because there were constraints at that time. The function of the division is to position Malaysia at multilateral discussion and formulating the policy. For implementation, the CCD facilitates climate finance, mitigation, and adaptation initiatives and strategies.

Researcher: In terms of the private sector's involvement, we wonder what ALAM's vision and objectives are.

Ms. Fatimah: ALAM's vision is to transform Malaysians' mindset and behaviours on the environment and its aim is to create awareness, educate people about climate change by collaborating with the public and private sectors as well as communities. ALAM also advocates policy for improved governance and bridging the policymakers with stakeholders.

Remarks

The respondents' responses revealed Malaysia is actively engaging in climate change policy and mitigation initiatives and strategies, at local and global level. Academicians contribute by providing data from research that can assist the government in formulating policies and developing mitigation and adaptation initiatives. The corporate sector also plays their role by engaging with the public to create awareness and promote actions.

As a researcher, the first respondent's insight into the climate change situation shows that Malaysia is experiencing unpredictable and extreme weather changes, like many countries around the world. With a temperature increase of approximately 1.1 degrees Celsius over the past 20 years, Malaysia is experiencing more extreme weather patterns, which includes heavy rainfall and drought. These changes contribute to issues like flooding and air pollution that affect the livelihood of farmers and fishermen, for instance. Thus, what are the government's actions or plans? Based on the other two respondents' responses, we can deduce that climate actions are largely driven by individual parties to fulfil their obligations under multilateral agreements such as *the Paris Agreement and the Kyoto Protocol.* Malaysia was the first Southeast Asian country that pledged its commitment to reducing emissions and building climate resilience. Meanwhile, national strategies like Long-Term Low Emission Development strategies (LT-LEDS) guide long-term goals where local initiatives are included in the broader green development plans, rather than sector-specific initiatives. In conclusion, Malaysia has made significant progress in addressing climate change through various policies and initiatives, like the National Climate Change Policy and Renewable Energy Policy, and continues to advance its climate efforts with the Malaysia Green Technology Agenda 2030. The policies and initiatives demonstrate the Malaysian government's commitment to tackling the worsening climate change issues. However, adaptation and execution of the plans is taking more time than expected.

Theme 2: Tensions Between Mitigation and Adaptation

The experts point out that there is an ongoing tension between climate change mitigation and adaptation efforts. Mitigation focuses on policies, and actions are aimed at handling climate change problems and issues, often driven by government initiatives and long-term goals. In contrast, adaptation involves changes in community behaviours and local responses to the immediate effects of climate change. While Malaysia has introduced several climate-related policies and the NGOs are also getting more actively involved, aligning these top-down mitigation strategies with adaptation remains a challenge. Bridging this gap requires greater coordination between the policymakers and communities to ensure both approaches support and reinforce one another.

Researcher: How do you define adaptation and mitigation? Would you say mitigation belongs to hard science? Or are they the same?

Dr Chee: Mitigation is when we are trying to do something—more to hard science. Adaptation is how people get used to it—more suitable for social science. Mitigation is more on the policy site and adaptation is more on the communitywise like behaviour. However, both are important.

Researcher: The government is good at making roadmaps, blueprint, etc., but implementation is where we are lacking. Also, with the political climate that we are having, perhaps that is one of the reasons that things are not moving to how it should be. What do you think about educating the public about climate change?

Dr Chee: It is a bit difficult because it involves so many things. Even educating children is a challenge because it needs to be interactive in order to attract them. Another problem for Malaysia is the energy part [sic]. Malaysia is an energy surplus country. We produce a lot of energy, so our energy is very cheap. For instance, oil and gas. So, changing to renewable energy will require more investment from the government [sic]. The transition might be a problem for the economy and society. Can people bear the cost? There is difficulty there because of the national energy profile [sic].

Ms. Fatimah: Yes, the government has launched a program; the National Adaptation Plan (MyNAP) in 2024 which will last until 2026. The objective of this program is to increase the country's endurance against the impact of climate change and to reduce long-term costs associated with climate change disasters. Additionally, the corporate sectors in Malaysia also have drawn up their own initiatives to help the government achieve this objective, which includes clean zero emission by 2050 such as energy efficiency and waste management.

Ms. Fatimah: Umm… as for the public, there is some effort and change in public behaviour in support of addressing climate change problems. For example, more people now carry reusable bags when shopping and with more recycling bins around, the public is also adopting recycling activities. Community garden is also another good and encouraging initiative carried out in neighbourhoods that is managed by the community to optimise the use of land and environmental space. These community gardens can increase community prosperity and create a sustainable environment.

Researcher: You also said that adaptation is significantly missing and going back to what you just said—that we're not challenging it. That is something very interesting and significant especially with the climate discourse and climate change situation now. I wonder why it is not taking place? Are we not aware? Are we not educated about climate change?

Madam Siti: There is a significant lack of capacity. The people who were involved with the negotiations in Paris (e.g., Tan Sri Zakri Hamid), the experts etc., are out of the system. That significant knowledge and exposure for multilateral discussion did not pass down to the successors. When you talk to the people of the industry, they have different views—they don't care about the policy since the market behaves differently from the policy. People don't necessarily have policy exposure. We don't really have good think tanks to provide us with good policy papers. Even in universities, when we talk about greenhouse gas emissions (GHG)—it's very engineering centric, it's not the same thinking. Malaysia really needs this pool of capacities with different backgrounds and perspectives and basically align to what policy and multilateral thinking is.

Ms. Fatimah: A survey on climate change awareness among Malaysians was conducted by the Monash Climate Change Communication Research Hub (MCCCRH) Malaysia in July— August 2022. The survey aimed to measure their awareness and concern about climate change as well as related behaviours in response to climate concerns. The report points out that not everyone is aware of the climate change that is happening, and it can be said that not many know what roles and actions they can take to address climate change issues.

Researcher: If I may add, I can consider myself as an environmental humanist, but I come from a literary background. I think in Malaysia we are not seen as important since we don't provide scientific and quantitative data. I think the government does not seek out beyond the scientific narrative about the environment. I think a lot of communication has been going on in other fields. I think you are right in the sense that we lack expertise but we also lack recognition of diversified areas that look at the environment. There's a need for paradigm shift, if I may suggest that.

Madam Siti: It's really disheartening to think that to be honest, most of the people who work under these sectors do not necessarily have the qualification (environmental background) to be there. There's a lack of insights on this (climate change) matters.

Ms. Fatimah: It is timely that collaboration between various parties (public and private) should be given more emphasis to increase public awareness. This awareness can be fostered from school by emphasising environmental protection elements in the school curriculum. This is what we at ALAM are focusing and working on. With the aim that emphasises "Mind-set transformation through education", we aim to educate Malaysians to respond to the related issues they encounter in their daily and working lives, and at the same time explore how to achieve change. We have been advocating this in various activities and programs such as World Sea Turtle Day, Coral Day, Environment Day, and others.

Researcher: Having said all these, we still have a long way to go in regards to educating the public about climate change. The government is actually doing something, not to say not doing anything, it's just that we are not there yet. We can't compare with those in the Western countries. However, is there any way that we as social scientists can go to the government and make them understand that they need to also listen to the narrative produced by us, social scientists and the public?

Madam Siti: Yes. Since we are pushing for development together with the climate change agenda. It has always been our principle when negotiating—we call it Common but Differentiated Responsibilities and Respective Capabilities (CBDR–RC) [a principle within the United Nations Framework Convention on Climate Change (UNFCCC) that acknowledges the different capabilities and differing responsibilities of individual countries in addressing climate change. Just to add, there is an internal project by Universiti Teknologi Mara (UiTM) at a village in Pahang to tackle the water source problem. They created an 8 km pipeline for the villagers but the maintenance is always an issue. However, the government is not aware of this matter. I personally think that in some cases, we need to talk to the people from the rural area about what they actually need. Sometimes they just want something that is operational. It is unlike what the government imagines [sic], it doesn't always have to be something big and require a lot of money. For instance, for flood mitigation, we spent about RM7 Billion during the Malaysian 11th Plan (RMK11) and then it went up to RM9 Billion. Now we are committing up to RM22 Billion until 2023. We, as a government, do not really talk about these small projects because the approach is always top-down. The government always thinks of big policies even though the public rather needs these small approaches to become climate resilient.

Remark

The interview clearly shows that Malaysia is constantly working on policies and plans that demonstrate the government's commitment to climate change mitigation, markedly prioritising greenhouse gas reduction and energy transition. However, the adaptation and implementation of these policies face significant hurdles. Many industry players are isolated from these governmental directives, as there is a dilemma for industry players to balance between market dynamics and policy frameworks. Concurrently, the general public lacks adequate exposure to these policies that impedes extensive understanding and engagement. This challenge is intensified by a shortage of specialists and intellectual groups who can formulate actionable policies that can address the chasm between government intent and real-world implementation. Despite these challenges and obstacles, various stakeholders are actively contributing to climate action. Private organisations are developing their own strategies and plans for emissions reduction and waste management. Meanwhile, the government needs to pay attention to initiatives from the public and private sectors because these smaller projects can contribute towards a larger climate resilient project, as mentioned by Madam Siti about the water woes problem in a small village in Pahang. Additionally, actions taken by the public such as active participation in recycling and community garden initiatives help foster environmental sustainability and also contribute to community prosperity.

Conclusion

Climate communication news discourse is shaped by political, economic, and cultural contexts. However, this news discourse often reflects global biases and Western dominance. A review of the literature on climate change and media reveals that media narratives vary across nations, with selective news reports and emphasis on the immediate or current climate change issues in that particular country. When climate change is framed as either urgent or distant, the media greatly shapes public understanding, engagement, and policy support. Eventually, the media plays a pivotal role in constructing climate change discourse, disseminating scientific knowledge about climate change and providing useful information in decision-making in government decisions on climate change issues. The Malaysian stakeholders' perspectives are also sought to understand their perspectives and insights into the issue of climate change. Two themes that emerge are policy implementation challenges and tension between mitigation, and adaptation efforts. The interviews show how climate change is discussed and acted upon in institutional, government, and corporate spheres, enriching the media analysis with voices from inside the academia, government policy, and corporate arenas. Despite the challenges and problems in handling increasing problems about climate change in the country, Malaysia is actively engaging in climate change policy and actions, at local and global level.

References

Azmi, N., Omar, N., Zaid, S., Wahab, Z., & Yusof, A. (2015). Media portrayal on global climate change An analysis of Malaysian mainstream newspapers. *Studies in Media and Communication, 3*(1), 73–78.

Barkemeyer, R., Figge, F., Hoepner, A., Holt, D., Kraak, J. M., & Yu, P. S. (2017). Media coverage of climate change: An international comparison. *Environment and Planning C: Politics and Space, 35*, 1029–1054. https://doi.org/10.1177/0263774X16680818

Carmichael, J. T., Brulle, R. J., & Huxster, J. K. (2017). The great divide: Understanding the role of media and other drivers of the partisan divide in public concern over climate change in the USA, 2001–2014. *Climatic Change, 141*(4), 599–612. https://doi.org/10.1007/s10584-017-1908-1

Carvalho, A. (2010). Media (ted) discourses and climate change: a focus on political subjectivity and (dis) engagement. *Wiley Interdisciplinary Reviews: Climate Change, 1*(2), 172–179.

Cheong, J. R., & Zanuddin, H. (2018). Media attention for climate change mitigation and adaptation in Malaysia: A comparative analysis of Malaysia Chinese newspaper coverage. Journal of Social Sciences Research, *Special Issue*(2), 413–419. https://doi.org/10.32861/jssr.spi2.413.419

Chinn, S., Hart, P. S., & Soroka, S. (2020). Politicization and Polarization in Climate Change News Content, 1985- 2017. *Science Communication, 42*(1), 112–129. https://doi.org/10.1177/1075547019900290

Eskjaer, M. (2013). The regional dimension: How regional media systems condition global climate-change communication. *Journal of International and Intercultural Communication, 6*(1), 61–81.

Fairclough, N. (1995). *Media discourse* (pp. 9–14). Edward Arnold.

Happer, C., & Philo, G. (2013). The role of the media in the construction of public belief and social change. *Journal of Social and Political Psychology, 1*(1), 321–336.

Happer, C., & Philo, G. (2016). New approaches to understanding the role of the news media in the formation of public attitudes and behaviours on climate change. *European Journal of Communication, 31*(2), 136–151.

Haris, M., Hassan, S., Tiwari, A. K., & Yao, H. (2025). *Impact of fiscal and green fiscal indicators on climate and environmental variability: the moderating role of technological innovation* (pp. 1–35). Environment, Development and Sustainability.

Hassan, I., Musa, R. M., Latiff Azmi, M. N., Razali Abdullah, M., & Yusoff, S. Z. (2024). Analysis of climate change disinformation across types, agents and media platforms. *Information Development, 40*(3), 504–516. https://doi.org/10.1177/02666669221148693$

Haris, S. M., Mustafa, F. B., & Ariffin, R. N. R. (2022). Exploring the Roles of Environmental Non-Governmental Organisations in the Context of

Malaysian Climate Change Governance. *International Journal of Sustainable Development & Planning, 17*(2).

Karkour, H. L. (2023). From the Twenty Years' Crisis to the climate crisis: Reconsidering Carr's thoughts on nationalism and global reform. *Journal of International Political Theory, 19*(3), 317–334. https://doi.org/10.1177/17550882231168906

Kim, K. (2011). Public Understanding of the Politics of Global Warming in the News Media: The Hostile Media Approach. *Public Understanding of Science.*

Kunelius, R., Eide, E., Tegelberg, M., Yagodin, D., & (Eds.). (2017). *Media and global climate knowledge.* Palgrave Macmillan US. https://doi.org/10.1057/978-1-137-52321-1

Konieczna, M., Mattis, K., Tsai, J. Y., Liang, X., & Dunwoody, S. (2014). Global journalism in decision-making moments: A case study of Canadian and American television coverage of the 2009 United Nations Framework Convention on Climate Change in Copenhagen. *Environmental Communication, 8*(4), 489–507. https://doi.org/10.1080/17524032.2014.909509

Manzo, K., & Padfield, R. (2016). Palm oil not polar bears: climate change and development in Malaysian media. *Transactions of the Institute of British Geographers, 41*(4), 460–476.

Martin, J. R., & White, P. R. (2005). *The Language of Evaluation: Appraisal in English.* Palgrave Macmillan Media Influence on Climate Change Perception. (n.d.). Elicit. Retrieved August 6, 2025, from https://elicit.com/review/fc104fc1-6d77-4f80-bb46-80088f8a3dd9

Mohamed, S., Yusoh, M. H., & Wan Ghazali, W. N. (2020). Framing sustainable energy: A comparative analysis of Malay and English newspapers in Malaysia. *International Journal of Asian Social Science, 10*(6), 295–306. https://doi.org/10.18488/journal.1.2020.106.295.306

Newman, T. P., Nisbet, E. C., & Nisbet, M. C. (2018). Climate change, cultural cognition, and media effects: Worldviews drive news selectivity, biased processing, and polarized attitudes. *Public Understanding of Science, 27*(8), 985–1002. https://doi.org/10.1177/0963662518801170

Newman, P., et al. (2020). *Resilient Cities: Responding to Peak Oil and Climate Change*. Island Press.

Olausson, U. (2009). Global warming—global responsibility? Media frames of collective action and scientific certainty. *Public Understanding of Science, 18*(4), 421–436. https://doi.org/10.1177/0963662507081242

Sampei, Y., & Aoyagi-Usui, M. (2009). Mass-media coverage, its influence on public awareness of climate-change issues, and implications for Japan's national campaign to reduce greenhouse gas emissions. *Global Environmental Change, 19*(2), 203–212.

Schmidt, A., Ivanova, A., & Schäfer, M. S. (2013). Media attention for climate change around the world: A comparative analysis of newspaper coverage in 27 countries. *Global Environmental Change, 23*(5), 1233–1248.

Sparks, C. (2007). Globalization, development and the mass media. SAGE Publications

Taufek, T. E., Nor, N. F. M., Jaludin, A., Tiun, S., & Aziz, J. (2022). Media Report About Climate Change in an English Online Malaysian Newspaper Through Thematic and Discourse Analysis Approaches. *3L. Southeast Asian Journal of English Language Studies, 28*(2).

Thaker, J., Zhao, X., & Leiserowitz, A. (2017). Media use and public perceptions of global warming in India. *Environmental Communication, 11*(3), 353–369. https://doi.org/10.1080/17524032.2016.1275733

Vu, H. T., Liu, Y., & Tran, D. V. (2019). Nationalizing a global phenomenon: A study of how the press in 45 countries and territories portrays climate change. *Global Environmental Change, 58*. https://doi.org/10.1016/j.gloenvcha.2019.101942

Weiner, R., Church, S. P., Lu, J., Esman, L. A., Getson, J. M., Fleckenstein, M., Radulski, B., Ranjan, P., Usher, E., Prokopy, L. S., & Pfeiffer, L. (2021). Climate change coverage in the United States media during the 2017 hurricane season: Implications for climate change communication. *Climatic Change, 164*(3–4). https://doi.org/10.1007/s10584-021-03032-0

2

A Multidimensional Approach to Climate Change Discourse

Introduction

This study examines climate change discourse in several selected mainstream and tabloid Malaysian newspapers. Following the interdisciplinary and data-driven nature of this project, a mixed-method approach encompassing corpus linguistics, sentiment analysis, Appraisal Theory, Critical Discourse analysis (CDA), and ecolinguistics analysis is employed. The selection of the analytical tools and theoretical bases is motivated by the need to provide both a general, empirically grounded overview of language patterns and a more interpretive exploration of their ideological implications. Data are sourced from two major English language newspapers in Malaysia, New Straits Times (NST) for mainstream perspective, and Malaysiakini for tabloid/alternative media outlets. This chapter describes the design, data collection, corpus construction, analytical framework, and data analysis procedure.

N. F. M. Nor et al., *A Critical Analysis of Climate Change Opinion News in Malaysian Media*, https://doi.org/10.1007/978-3-032-10333-8_2

Background to the Multidimensional Approach

The multidimensional approach employed in this study is consistent with Creswell and Plano Clark's (2018) typology of convergent parallel approaches. Incorporating corpus-driven techniques and qualitative discourse analysis, this approach draws on both computational strengths of corpus tools and interpretive depth of qualitative textual reading. The nature of climate change communication, which is complex and often ideologically fraught, demands such a hybrid approach. We believe the multidimensional approach can offer a multi-perspectival lens through which surface-level textual patterns can be related to more socio-discursive insights.

In particular, the approach was chosen because it allowed the study to achieve its dual goals: (1) recognising recurring linguistic patterns in public opinion news on Malaysian climate discourse, (2) decoding associated attitudinal, emotional, and ideological positions, and (3) uncovering the ecological ideas and ideologies in our lives. Corpus analysis allows robust generalisation to large dataset, sentiment analysis, and critical discourse analysis allow for micro-level interpretations of evaluative stances, and ecolinguistics reveal ecological ideas and ideologies embedded in communication, and how these linguistic constructions influence perceptions, behaviours, and policies related to the environment. These are not treated as the only methods but are interwoven through triangulated interpretation. To recapitulate, the sentiment lexicons generated during sentiment analysis are used to select concordance lines for further analysis using Appraisal Theory at the sentence level. Those selected evaluative expressions are then used as the basis for further interpretive analysis through critical discourse analysis and ecolinguistics analysis.

The use of representative textual genres is added as one of its features or parameters that focuses only on opinion news that express public and editorial attitudes—e.g., letters to the editor, commentaries, and op-eds. To reiterate, opinion news is news that serves to explicitly demonstrate public opinion/sentiment on a certain issue. It is written by citizens who are not directly associated with the news organisation and can be used to informally measure public opinion (Brooker & Schafer, 2015). In selecting such texts, a constructivist account of language was employed in which the media are treated as active shapers of public understanding and not as conveyors of fact. Hence, this allows for a description of media discourse and how power, ideology, and affect operate within it. A systematic comparison across media types—the dichotomy between mainstream and tabloid/alternative—was selected because these comparisons provide an important insight into how climate change is discussed, and how different editorial contexts might affect evaluative stance, topic prominence, and ideological framing (Grantham & Vieira Jr., 2014). The multidimensional approach and sampling across context mean that the approach is not dependent on a single method, hence, more data source is included to increase the reliability and quality of the findings.

Climate Change Corpus

These data are taken from English language opinion news published between 2015 and 2021 regarding climate change and related environmental discourse. The seven years data coincided with a few reasons: year 2015 was chosen based on the adoption of the Paris Agreement by the parties of UNFCCC, including Malaysia (United Nations Climate Change, n.d.), while year 2021 was chosen due to a significant event in Malaysia's climate change history—a major flash flood that affected more than 5 states, including Selangor, an area not typically prone to flooding (Cameons, 2021). This has shown an increase in public discourse on many climate-related discussions. By selecting this timeframe, the corpus captures a dynamic policy and media landscape reflecting political intent as well as societal concern.

Two newspapers had been chosen due to their prominence, ideological range, and accessibility: New Straits Times represented the mainstream press while Malaysiakini represented the alternative/tabloid outlets with more independent, sensational, or populist editorial angles (Boykoff & Mansfield, 2008). Such stratification opens up a comparative study of discourse constructions across institutional contexts and reveals more about how ideologies and communicative strategies differ across perceived media affiliations.

Articles were retrieved manually from digital archives and online repositories using predefined search terms: "climate change Malaysia", and "climate change". Pilot testing of these search terms produced relevant opinion news while reducing irrelevant results. To avoid inconsistent reporting and analytical rigour, only articles clearly labelled as opinion and those that only talk about climate change or environmental issues as the main subject were included.

There were 245 opinion news articles selected and curated for analysis, comprising 135 articles from Malaysiakini and 110 articles from the New Straits Times. HTML elements, advertisements, author biographies and unrelated hyperlinks were removed from each article to limit the data to the editorial content only. The news articles were saved in plain.txt format with standardised naming convention including publication name, article date, and document ID. This was done to assist in better organisation of the date and to allow the data synchronisation into the tool that we used for analysis which was *AntConc*.

Multidimensional Approach Design

The analytical framework utilised in this study is intentionally multilayered; this is needed for us to move from broad descriptive overviews to fine interpretive insights. This approach reflects a growing recognition in discourse and media studies that no single method can account for climate change communication complexity (Moser, 2010; Carvalho & Burgess, 2005). By sequentially layering corpus linguistics, sentiment analysis, Appraisal Theory, Critical Discourse Analysis, and ecolinguistics, it captures how climate change is framed, felt, and ideologically constructed in the Malaysian media.

Simply put, the approaches are selected because they complement each other and are crucial in creating an observable matrix. Corpus linguistics offers reproducible scalable methods for detecting large lexical patterns in a large textual dataset. Sentiment analysis adds an emotional dimension, measuring affective polarity based on how issues are perceived and internalised. On the other hand, through Appraisal Theory, these emotional and evaluative positions can be classified and interpreted. Critical Discourse Analysis allows for a closer examination of the power relations and ideological work of the text. Ecolinguistics is applicable in exploring general patterns of the world that influence how people think and treat the world. Resulting from each method is a build on the previous one, an interpretive chain linking the empirical with the critical.

This layered approach is consistent with the study goals as it would be useful to reveal not just what is said about climate change in the media, but how it is said, with what evaluative tone, and to what potential social and ideological effect. These integrative approaches allow for breadth and depth, whereby breadth is through the corpus lens and depth is through sentiment, appraisal, and critical discourse interpretation. The multidimensional approach enables the analysis to probe deeper into the socio-political dynamics underpinning climate change discourse in Malaysia.

To recapitulate, the analysis uses five related approaches. The first is corpus linguistics which is a descriptive overview of language patterns. The second is sentiment analysis which is used for evaluating emotional polarity. The third is Appraisal Theory, an analysis of interpersonal meaning. Fourth, Critical Discourse Analysis (CDA) explores power, ideology, and social relations embedded in climate discourse. Finally, ecolinguistics framework, which is used to analyse the relationship between language and the environment.

Corpus Linguistics and Environmental Discourse

Corpus linguistics has emerged as a methodological and theoretical basis for the empirical study of language, allowing for pattern recognition in large corpora of naturally occurring text (Baker, 2010; McEnery & Hardie, 2012). A corpus is a principled, electronic set of authentic texts that permits researchers to go beyond introspection or isolated examples to observe and quantify linguistic phenomena. For climate change discourse, this is particularly relevant because it allows a critical examination of how the issue is represented, contested, or communicated in the media, policy documents, or public opinion texts over time and across genres (Stubbs, 1996).

In addition, corpus linguistics research requires a combination of quantitative and qualitative approaches to language. Statistically, corpora provide frequency counts, collocational patterns, and statistical associations (e.g., keyness, concordances) showing how specific words, phrases or grammatical features act in context. Qualitatively, these patterns may then be interpreted to reveal deeper discursive strategies, evaluative positions, or ideologies underlying language use (Partington et al., 2013). Without both, the findings are or feel truncated.

One central argument why corpus linguistics is employed because it has a particular value that preserves linguistic authenticity, as McEnery and Hardie (2012) explain. For example, the texts reflect real-world language, thus enhancing the ecological validity of findings. Furthermore, it is well suited to discerning what is said and not said—the silences and omissions or obfuscations that shape meaning implicitly (Baker & McEnery, 2015).

For over two decades, corpus linguistics has been applied widely to analyse media discourse, particularly environmental and socio-political discourse. It permits the analysis of both broad patterns and specific details (Marchi, 2022). For instance, Baker et al. (2013) analysed Muslim representations in the British press using 143 million words to show how some collocates constructed Muslims as threatening or victimised. Similarly, in climate change research, corpus approaches have been used to map denialist frames, moral evaluations, and urgency narratives in news media (Flottum et al., 2016; Grundmann & Krishnamurthy, 2010).

These studies demonstrate that corpus linguistics is appropriate for examining patterns at scale, patterns that escape researchers after skimming a few articles, and for detecting subtle evaluative or ideological tendencies in discourse. In a similar vein, we utilise Malaysian opinion news on climate change as, and corpus linguistics is applied to examine how climate issues are presented in different newspapers, how moral responsibility is assigned and how urgency versus sustainability narratives emerge.

Another reason why corpus linguistics fits this book is its complementarity with critical discourse analysis (CDA). While CDA seeks to reveal the ideological underpinnings of discourse and its role in maintaining social inequalities (Fairclough, 1995; van Dijk, 1998), corpus linguistics strengthens these insights by adding empirical, quantitative evidence. Baker (2006) describes this as a corpus-assisted discourse studies (CADS) approach, which combines the explanatory power of CDA with the replicability and breadth of corpus methods.

As an illustration, in this book, sentiment lexicons are used to select relevant texts. Then, they examined through concordance analysis to categorise evaluative stances at the sentence level. Corpus evidence identifies key attitudes and discourses, and gives not only a bird's-eye view of the data but also allows a minute examination to study how language constructs meanings and positions readers.

We would argue that corpus linguistics inclusion is methodologically sound and necessary for a few reasons. First, it is scalable and robust. Climate change discourse has thousands of texts and corpus methods that allow for a systematic rather than a selective analysis of the materials. Second, it also reveals discourse patterns and recurring frames that are not immediately apparent from close reading alone. Indeed, this allows a more nuanced reading and interpretation of the data (Ilyinova & Kochetova, 2024). Third, a mixed-method approach is another strength, as corpus linguistics combines quantitative rigour with qualitative interpretation, thereby demonstrating the value of integrating multiple methods within a single analytical framework. Therefore, corpus methods make findings transparent and reproducible. Finally, this approach fits global trends in climate communication research that are increasingly computational and corpus-based, placing this study within and contributing to this scholarly conversation.

In short, corpus linguistics provides a complementary methodology to sentiment analysis, Appraisal Theory, CDA, and ecolinguistics, to discover how climate change is communicated in Malaysian opinion news. It lets researchers answer questions about what is said, how it is said and what is left unsaid with empirical data.

Sentiment Analysis of Public Opinion News

Sentiment analysis is a computational approach to identifying, extracting, and quantifying subjective information from text (Liu, 2012). It is also called opinion mining. For over two decades, it has been a core technique of natural language processing (NLP), allowing researchers to evaluate affective orientation (positive, negative, or neutral) of large-scale textual data. The premise of sentiment analysis is that language not only conveys information factually but also conveys attitudes, emotions, and evaluations (Taboada et al., 2011).

The conceptual basis in sentiment analysis rests on the assumption that text often codes evaluative meaning in terms of lexical and grammatical choices. For example, adjectives like *remarkable* or *disastrous*, modal verbs like *obligation* or *doubt*, and intensifiers all contribute to the overall sentiment of a sentence or document (Pang & Lee, 2008). Often these cues are detected computationally via lexicons (lists of words and phrases labelled with the typical sentiment orientation) or machine learning models trained on annotated corpora (Cambria et al., 2017).

Lexicon-based approaches are therefore ideally suited for studies aiming at linking observed sentiment to specific linguistic choices. Machine learning methods are often more accurate on benchmark tasks but may be "black boxes" (a system where inputs and outputs are visible but internal decision-making is opaque or hard to interpret) and require large training datasets which are not always available for specialised domains like climate change discourse. In this study, lexicon-based sentiment analysis was applied to identify and extract evaluative language from Malaysian opinion news as a structured way to isolate texts with high affective content for further appraisal and discourse analysis.

In media studies, sentiment analysis is commonly applied to monitor public attitudes and media stances towards events, policies, and social issues. Pang and Lee (2008) described its use for analysing product reviews, political speeches, and online debates, while Thelwall et al. (2010) showed its application to tracking emotional trends in social media data. Specifically for climate change communication, sentiment analysis has examined how media outlets frame the urgency, severity, and moral implications of climate change. For example, O'Neill et al.

(2015) used sentiment analysis to assess the balance of hope versus fear in climate news, while Schmidt et al. (2013) measured changes in polarity of climate-related reporting over time. These studies illustrate how sentiment analysis reveals evaluative dimensions of discourse at scale. Since opinions often serve as platforms for author opinions and judgements about climate change policies, actors, and consequences, sentiment analysis is ideally suited for finding such patterns in the corpus.

The inclusion of sentiment analysis reflects its aim to provide empirical insights into the evaluative and ideological dynamics of climate change discourse in Malaysia. By applying sentiment lexicons to a large corpus of opinion news, the study was able to detect which texts exhibit strong subjective language, positive or negative, and which lean towards neutrality. This served not only as a diagnostic tool but also as a selection criterion for subsequent appraisal and critical discourse analysis.

Sentiment analysis complements corpus linguistics by quantifying evaluative stances which might be scattered or obscured in large datasets. Its application allows an easily replicable and systematic way of locating sentiment-laden discourse within broader communicative patterns. This shows for readers of this book—researchers and practitioners interested in climate communication—that computational methods may augment qualitative methods in providing actionable insights into affective and ideological dimensions of environmental messaging.

Therefore, sentiment analysis constitutes a useful methodological interface between the quantitative breadth of corpus linguistics and the qualitative depth of appraisal and critical discourse analysis. This allows researchers to systematically surface subjective and evaluative language for text selection and interpretation. In this book, sentiment analysis is crucial to understand how Malaysian media frame climate change—patterns of blame, responsibility, urgency, and hope embedded in public discourse.

Climate Change Attitude through Appraisal Theory

Martin and White (2005) developed Appraisal Theory based on Halliday's Systemic Functional Linguistics (SFL) and described how language conveys attitudes, negotiates social relationships and positions readers or listeners. As an extension of the interpersonal metafunction of language, it focuses on the evaluative dimension of discourse: how speakers and writers express feelings, assign value, and align with or against others. This is particularly appropriate for considering ideological and affective aspects of climate change communication, where issues of blame, responsibility, and urgency are central.

The appraisal framework has three interrelated subsystems: attitude, engagement, and graduation. Attitude involves feelings and values expressed in affect, judgement, and appreciation. Affect includes emotional reactions such as fear, hope or anger, and judgement includes moral or ethical evaluations of people's behaviour, such as calling someone responsible, negligent, or visionary. The aesthetic or evaluative appraisal of things, processes or phenomena is called appreciation if an event is significant, harmful, or remarkable. Those three subsystems together define the variety of evaluative meanings that writers may encode for readers to affect their evaluations of people, actions, and events. The engagement system concerns how writers recognise and respond to other voices and perspectives within discourse, and graduation concerns the degree of intensity, amplification, or mitigation of evaluations. Together, these subsystems form a lens for analysing evaluative meanings in text.

The Appraisal Theory has been used in media and communication research to discover how ideology and stance are communicated through language. For instance, White (2008) demonstrated that evaluative language in Australian newspapers constructed particular ideological positions about asylum seekers; whereas, Bednarek (2006) investigated how newsworthiness was linguistically encoded through evaluative resources in broadcast media. In the environmental context, Stibbe

(2015) investigated how language reflects and shapes human relationships with nature, showing how evaluative choices promote or undermine ecological awareness. These studies demonstrate how Appraisal Theory allows researchers to identify not only whether language is positive or negative, but also what kinds of evaluations are made, towards whom or what, and how strongly those stances are expressed. This is especially useful in the context of climate change to understand how media discourses mobilise emotions, frame responsibility, and construct moral imperatives.

In this study, Appraisal Theory is applied mainly to the attitude sub-system since opinion news, as soft news, inevitably contain affective and moral language as they often hinge on personal inclinations, faith, ideology, and values (Heinderyckx, 2021). The sentiment analysis identifies texts which contain strong subjective language which are then analysed more closely through the appraisal lens to classify the evaluative stances present. This permits wide interpretation of how opinion writers assign blame, assign moral or ethical responsibility, and frame climate change as urgent or manageable. The gap between the quantitative insights of sentiment analysis and critical discourse interpretation can thus be plugged in Appraisal Theory.

Appraisal Theory also offers a linguistically grounded yet flexible framework for revealing how Malaysian opinion news position readers and construct evaluative meanings about climate change. This framework reflects the interdisciplinary nature of the study which combines corpus linguistics, sentiment analysis, and Critical Discourse Analysis. A detailed taxonomy of evaluative resources aids Appraisal Theory in making the findings transparent and interpretable, showing precisely how ideological and affective positions are encoded in language. It also elaborates micro-linguistic strategies contributing to broader discursive and ideological formations in addition to critical discourse analysis.

Finally, Appraisal Theory provides a system for exploring how writers feel, judge actions, and evaluate phenomena in discourse. It enables researchers to move beyond general sentiment labels towards a more moral, emotional, and aesthetic conception of language. The Appraisal Theory considered herein provides important insights into how

Malaysian media construct urgency, blame, and hope in public perceptions of climate change, and is therefore an indispensable tool in the methodological toolkit.

Critical Discourse Analysis

Critical Discourse Analysis (CDA) has evolved from the Critical Linguistics (CL) movement, with the latter being developed in the 1970s, primarily from the University of East Anglia, aiming to expose how language in texts, especially media encoded and reproduced ideology and power relations. Its focus was on linguistic features to reveal underlying biases. However, CL was limited in its scope, with the focus on examining language at textual level of analysis without adequately connecting linguistic choices to broader social contexts and practices. Consequently, this led to the development of CDA, which evolved to overcome CL limitations. CDA recognises that discourse should not just analyse language but also consider language as a form of social practice that both shapes and is shaped by society. CDA aims to analyse how power abuse, dominance, and inequality are enacted, reproduced, and resisted in text and talk. In summary, CDA goes beyond linguistic analysis or structures to incorporate social, historical, and political contexts; it also incorporates other disciplines such as sociology, critical theory, philosophy, and politics, hence interdisciplinary in nature. Indeed, it views language as a tool for social action and reformation, and not just a reflection of ideology, and contributes to social change by uncovering how power and inequality is embedded in texts and talk. Therefore, CDA is a general approach used to study any form of discourse related to social issues and is regarded as an approach that investigates discourse by analysing the interrelationships between language, power, and ideology.

van Dijk (in Schiffrin et al., 2015) points out that CDA aims to uncover the covert power dynamics and ideological underpinnings that are inherent in language use and to contest dominant discourses and representations that might perpetuate social inequities. The term CDA is used nowadays to refer more specifically to "the critical linguistic approach of scholars who find the larger discursive unit of text to be the basic unit of communication" (Wodak, 2001, p. 2). Fairclough (1995) defines CDA as a form of discourse analysis which aims to systematically explore often opaque relationships of causality and determination between (a) discursive practices, events and texts, and (b) wider social and cultural structures, relations and processes. Concomitantly, CDA is employed to investigate how such practices, events, and texts arise out of, and are ideologically shaped by the relations of power and struggles over power; and to explore how the opacity of these relationships between discourse and society is itself a factor securing power and hegemony. To put it simply, CDA aims at making obvious the connections between discourse practices, social practices, and social structures; connections that might not be clear for a person unfamiliar with the field. Consequently, three concepts are found in all CDA approaches: the concept of power, the social practices, and social structures.

van Djik (2015) further emphasises that CDA acts as a means to challenge the status quo and effect social change by exposing how language use can be wielded to maintain or transform existing power structures. As mentioned by Van Leeuwen (in Wodak & Meyer, 2009), the primary goal of CDA is to demonstrate how this process occurs. It also aims to focus on specific social issues and understand them by analysing the discourse related to the issues in question, which are produced by socially situated speakers and writers. Therefore, the idea of discourse is defined as a social operation in which ideologies can be both shown and produced. In CDA, the concept of discourse encompasses both the social conditions of production (e.g., text) and the social conditions of interpretation. The text is the realisation of the environment in which it functions through the process of interaction and semantic choice (ibid).

Additionally, CDA operates on the belief that language is not impartial but rather moulded by social, political, and historical situations (Wodak & Meyer, 2009). An examination of how language is used in different types of communication, including media, politics, education, and daily interactions uncovers how power dynamics are established, upheld, and challenged through language usage (Fairclough, 2013). In essence, CDA is well suited for scrutinising and contesting how language use perpetuates and sustains social disparities. As discussed earlier, is an interdisciplinary approach that incorporates perspectives and techniques from various disciplines such as linguistics, sociology, psychology, anthropology, among others, to offer a comprehension of the intricate relationships between language use and social reality (Blommaert & Bulcaen, 2000). By revealing the covert power structures and ideological elements in communication, CDA can aid in advancing social equity and impartiality (Wodak, 2009).

Critical Discourse in Media and Climate Discourse

CDA plays an essential role in revealing hidden ideologies and examining the presence of power in media discourse studies. Ramanathan and Hoon (2015) state that CDA is a discourse analysis branch that goes into how and why discourse reproduces macro systems cumulatively and highlights traces of cultural and ideological importance. In the debate about the term CDA, the word 'critical' is discussed in various ways to understand the use of language and the reproduction of power in culture. Critical is clearly associated with the issue of control, superiority, and resistance in the different language fields. Fowler (1991: p. 23) asserts that "newspapers in part adopt this language for their own and, in deploying it, reproduce the attitudes of the powerful group". In news discourse, CDA strives to examine how language and discourse are employed to contrive and sustain dominant news frames and narratives, and to recognise how news can serve as a form of opposition and societal transformation. In fact, CDA is a theoretical and analytical structure that attempts to investigate how language and discourse are employed to shape power dynamics and maintain existing ideologies.

News discourse refers to the discourse spread, used, or manufactured in news media, raising concerns about the relationship between language and ideology (Fairclough, 1992). Halliday (2007) proposes that language does not passively reflect reality, but actively constructs reality. Therefore, the CDA approach is suitable to analyse the multilayered language of newspapers and uncover underlying biases created by journalists to find out if the texts implied to sustain power or construct reality in a desired manner by journalists or ideological groups (Sajid et al. 2019). Language used in journalistic discourse is of significant importance in our society that operates at social and linguistic level. This implies that journalism and language are both social instruments that serve specific functions in the social system in which they are used. According to Richardson (2017), journalistic discourse has specific textual features that are closely linked to a CDA framework. Richardson identifies three key aspects that characterise journalism: language, production and consumption, and its relations to social ideas and institutions, claiming that the way news is obtained and created is closely related to people's opinions, making it a crucial area for critical analysis. Consequently, the social issue in question should be presented from the viewpoint of those who suffer the most to criticise those in power, those responsible, and those who have the means to solve such social issues (van Dijk 2001, as cited in Richardson 2017). Richarson (ibid), however, cautions that although journalism exists to enable citizens to better comprehend their lives and position in the world, it may be challenging to determine whether journalism can truly provide us with a comprehensive understanding of the world, considering the presence of bias in news reports.

The concept of ideology exists in various forms, and although some scholars tend to focus on aspects of "illusion", "false consciousness", and "fetish" when describing it, these are not the primary features of ideologies. van Dijk (1998) points out that ideologies enable individuals to organise their beliefs about what is right or wrong, good or bad, and to act accordingly as members of a group. Essentially, ideologies provide the foundations for shared representations among social groups. Building on this understanding, several past studies have applied CDA to examine how ideologies and power relations are presented in climate change discourse, particularly in news media. The reviewed past studies (e.g. Abdellah, 2020; Bureau, 2024; Carvalho, 2007; Garlitos, 2020; Huang and Che, 2024; Orhewere & Olley, 2023; Pepermans & Maeseele, 2018; Wagner & Payne, 2017; Ye et al. 2024) reveal how language is manipulated and strategically used to position institutional interests and ideological positions which influence the framing of climate change issues. Newspapers vary from national newspapers in the UK, USA, China, and Belgium to local dailies and UN news. Additionally, online news comment sections, media websites, and climate change conference reports were used as data. These studies used CDA, notably Fairclough's three-dimensional framework (1995), in addition to van Djik's (1993) socio cognitive approach to CDA, Wodak 's (2001) Discourse Historical approach (DHA), and Halliday's (1967, 1978, 1985) Systemic Functional Linguistics (SFL), and corpus linguistics approach are also employed to uncover how newspapers used language to present ideology and power structures in climate change news discourse. These findings detail how media discourse on climate change not only reflects national and elite ideologies but also actively constructs power structures through language.

With the growing number of environmental concerns that the Philippines has been facing, Andrew and Garlitos (2020) examined 25 mainstream news reports on environmental concerns using Fairclough's (1992) CDA, and Halliday's (1985) Systemic Functional Linguistics (SFL). The results reveal that the themes in the news reports represent nature as the enemy and the culprit of destruction, while the government are presented as the eco-warriors. Interestingly, the public are seen as weak and defenceless against the authorities who are viewed as empowered and influential. Additionally, plants and animals are seen as human commodities, further emphasising the creation of hierarchical power discourse. Andrew and Garlitos (ibid) conclude that the results clearly demonstrate how linguistic patterns are implicitly and explicitly manipulated to serve and promote ideological assumptions of the news. They illustrate how the writers' choice-making is plagued with biases since they decide what to include and exclude in the news discourse.

A study by Huang and Che (2024) explores the linguistic features of climate change news discourse in China and the United States, the discursive processes and possible social factors that are used to strengthen both countries' national ideologies. The research corpus comprising 128 reports from the *China Daily* and *Los Angeles Times* from November 18, 2022, to January 4, 2023, were collected within 2 months after the Conference of the Parties to the United Nations Framework Convention on Climate Change (COP27). They used corpus linguistics (CL) approach and Fairclough's (1995) Three-Dimensional CDA framework. The study reveals that both Chinese and American news focus on climate issues and energy industry development around keywords such as *climate*, *global*, *energy*, and *emissions*, and call for countries around the world to participate in environmental protection. These types of news appear to consolidate national ideologies of both countries.

Similarly, the American media highlights the leadership of American climate governance through discourse, emphasising the universal value of democracy and strengthening the national identity of the American people. Despite the similarities, differences prevail between Chinese and American climate news that are prominently reflected in the social origins of the linguistic features. Since the mission of *China Daily* is to promote communication between China and the world, the newspaper often emphasises China's environmental actions and international cooperation when communication difficulties happen. On the other hand, The *Los Angeles Times* is concerned about sales of the newspaper, hence, controversial topics are frequently portrayed. Discourse strategies to consolidate national ideologies in the climate news discourse of both countries are mainly through the concentrated reporting on climate issues like industrial development, emission issues, and government actions, to clarify climate responsibility, and shape a responsible national image that is actively solving climate problems.

The news media in China and the United States attach great importance to reflecting objective attitudes in reporting and use modal verbs of median value to reduce reporters' responsibilities. However, there is no complete objectivity and neutrality in news reporting (Fowler, 2013). The Chinese media hopes to promote public recognition of China's climate policies and national pride by showing the country's willingness to cooperate with other countries. This is in line with previous findings of Chinese reporting, which tends to highlight positive developments and requires global support to tackle climate change (Song et al., 2021). The American media applies a similar strategy by stressing on shaping the leadership of American climate governance through discourse, conveying the universal value of democracy, and strengthening the national identity of the American people.

Ecolinguistics in Climate Change Discourse

Language influences how we think about the world, thus, our thoughts, concepts, ideas, ideologies, and worldviews are shaped through language. The link between ecology and language is how humans treat each other and the natural world. Additionally, how we think has an influence on how we act, which means that language can inspire us to destroy or protect the ecosystems that life depends on. The term 'ecolinguistics' has been used since at least in the 1990s (Fill & Mühlhäusler 2001: 1), and the word 'ecology' has been incorporated in linguistic accounts since (at least) Einar Haugen's work in the 1970s (e.g. Haugen 1972). Furthermore, the term 'ecolinguistics' has been applied to a wide range of approaches and interests, some more relevant than others to the larger 'ecological turn' in humanities and social sciences. The term has been used in studies on language interaction and diversity; such as outdoors signposts, text analysis about the environment, how language relates to objects in the local environment, studies of dialects in particular geographical locations, and many other diverse areas (Stibbe, 2015). Based on the different approaches, Steffensen and Fill (2014: 7) iidentify four different interpretations of ecology. The first approach views language as existing in a symbolic ecology, where different languages interact with each other in a given location. The second approach sees language as part of a sociocultural ecology where it shapes societies and cultures. The third approach is concerned with cognitive ecology and how the cognitive capacity of organisms affects how they adapt to their environment. Finally, there is a natural ecology which is concerned with the relationship of language to its biological and physical environment. It is within the four approaches that ecolinguistics plays a crucial role. As Stibbe (2015, p. 1) points out, ecolinguistics is about critiquing forms of language that contribute to ecological destruction, but also aids in the search for new forms of language that inspire people to protect the natural world". Stibbe also (ibid) emphasises how ecolinguistics is applicable in exploring general patterns of the world and influences how people think and treat the world. He refers to this idea as the "stories we live by, i.e. mental modes that influence behaviour and lie at the heart of ecological changes that we are facing" (pp. 1–2). The concept of ecolinguistics can be used to examine ideologies, metaphors, frames, and a variety of other cognitive and linguistic phenomena through the process of revealing and uncovering the stories that shape people's lives and shape the society in which we live. The ecological turn has given rise to environmental communication which Milstein et al. (2009: 344) summarises as follows:

> Research and theory within the field are united by the topical focus on communication and human relations with the environment. Scholars who study environmental communication are particularly concerned with the ways people communicate about the natural world because they believe such communication has far-reaching effects at a time of largely human-caused crises.

Stibbe's (2015) proposes an ecolinguistics framework which consists of eight categories as depicted in Table 2.1.

Table 2.1 Stibbe's (2015) Ecolinguistics Framework

Category	Description	Language elements
Stories we live by	What fundamental beliefs or narratives about the environment are being promoted?	Humans as conquerors of nature, nature as a resource.
Framing	The use of source frame to structure a target domain	Trigger words that bring a particular source frame to mind
Ideology	A story about how the world was, is, and should be in the minds of the members of the group	Discourse, i.e. characteristics of language features used by members of the group.
Metaphor	A type of framing where the source frame is from an imaginable area of life that is distinctly different from the target domain	Trigger words that bring the source frame to mind
Evaluation	A story in people's minds about whether an area of life is good or bad	Appraisal patterns, i.e. patterns of language which represent things positively or negatively.
Identity	A story in people's minds about what it means to be a particular kind of person	Forms of language that characterise people
Conviction	A story in people's minds about whether a particular description is true, certain, uncertain, or false	Facticity patterns, i.e. linguistic patterns that represent a description as true, uncertain, or false
Erasure	A story in people's minds that something is unimportant or unworthy of consideration	Patterns of language which erase or diminish
Salience	A story that something is worthy of consideration	Patterns of language which foreground an area of life

Conclusion

This chapter describes and discusses the approaches used for analysing opinion news on climate change. The chapter explains in detail the process of filtering and selecting opinion news on climate change in Malaysia which leads to the development of a specialised corpus, known as the Malaysian Climate Change Corpus (MyCCC). A hybrid approach is deemed necessary for analysing opinion news on climate change because climate change discourse is linked to various aspects such as economy, social and cultural, and is often embedded with ideology. This method combines different perspectives to connect surface-level language patterns with deeper social and cultural meanings. The corpus linguistics approach enables generalisations across large datasets, while Appraisal theory, critical discourse analysis, and ecolinguistics framework provide detailed micro-level interpretations of the discourse. Appraisal Theory is a useful framework for analysing attitudes on climate change in opinion news because the analysis delves into the specific values, emotions, and stances expressed through language. By examining a text's evaluative language, Appraisal Theory can reveal the underlying ideologies and positions of authors and news outlets, which provides insights into an individual's perceptions and attitudes on climate change. Additionally, the application of critical discourse analysis (CDA) for analysing attitudes on climate change in opinion news goes beyond the surface level of text to examine how language is used to construct power, ideology, and social reality. CDA examines how a text is constructed and why it is constructed, hence, revealing the underlying attitudes and agendas of the authors and institutions. Finally, the ecolinguistics framework reveals how language constructs the "stories" that shape our relationship with the environment because it focuses on how language can either reinforce harmful ecological attitudes or raise sustainable attitudes. Therefore, this integrated approach ensures that the study not only generalises from large datasets, but also facilitates an understanding of individual expressions, as expressed in opinions about climate change.

References

Abdellah, A. (2020). The islamic declaration on global climate change; an ideological discourse analysis. *Middle Eastern Journal of Research in Education and Social Sciences, 1*(2), 77–93.

Baker, P. (2006). *Using corpora in discourse analysis.* Continuum.

Baker, P. (2010). *Sociolinguistics and corpus linguistics*. Edinburgh University Press.
Baker, P., Gabrielatos, C., & McEnery, T. (2013). *Discourse analysis and media attitudes: The representation of Islam in the British press*. Cambridge University Press.
Baker, P., & McEnery, T. (2015). *Corpora and discourse studies: Integrating discourse and corpora*. Palgrave Macmillan.
Bednarek, M. (2008). *Evaluation in media discourse: Analysis of a newspaper corpus*. Continuum.
Blommaert, J., & Bulcaen, C. (2000). Critical discourse analysis. *Annual Review of Anthropology, 29*(1), 447–466.
Boykoff, M. T., & Mansfield, M. (2008). "Ye olde hot Aire": Reporting on human contributions to climate change in the UK tabloid press. *Environmental Research Letters., 3*(2). https://doi.org/10.1088/1748-9326/3/2/024002
Brooker, R. G., & Schaefer, T. 2015. Methods of measuring public opinion. Public opinion in the 21st century. https://www.uky.edu/AS/PoliSci/Peffley/pdf/473Measuring%20Public%20Opinion.pdf
Bureau, P. (2024). Climate knowledge or climate debate? Using word embeddings and critical discourse analysis to compare expert and media representations of climate knowledge. *Terminology, 30*(1), 35–57.
Cambria, E., Schuller, B., Xia, Y., & Havasi, C. (2017). New avenues in opinion mining and sentiment analysis. *IEEE Intelligent Systems, 28*(2), 15–21.
Cameons, A. (2021). More than 1,000 evacuated from flood-hit areas in Klang. *The Star*. https://www.thestar.com.my/news/nation/2021/12/18/more-than-1000-evacuated-from-flood-hit-areas-in-klang [10 January 2022]
Carvalho, A., & Burgess, J. (2005). Cultural circuits of climate change in UK broadsheet newspapers, 1985–2003. *Risk Analysis: An International Journal, 25*(6), 1457–1469.
Carvalho, A. (2007). Ideological cultures and media discourses on scientific knowledge: Re-reading news on climate change. *Public Understanding of Science, 16*(2), 223–243.
Creswell, J. W., & Plano Clark, V. L. (2018). *Designing and conducting mixed methods research* (3rd ed.). SAGE Publications.
van Dijk, T. A. (1993). Principles of critical discourse analysis. *Discourse & Society, 4*(2), 249–283. https://doi.org/10.1177/0957926593004002006
van Dijk, T. A. (1998). Ideology: A multidisciplinary approach.
van Dijk, T. A. (2001). Discourse, ideology and context.

van Dijk, T. A. (2015). Critical discourse analysis. In D. Schiffrin, D. Tannen, & H. E. Hamilton (Eds.), *The handbook of discourse analysis* (2nd ed.). Wiley-Blackwell.

Fairclough, N. (1992). *Discourse and social change*. Polity Press.

Fairclough, N. (1995). *Critical discourse analysis*. Longman.

Fairclough, N. (2013). *Critical discourse analysis: The critical study of language*. Routledge.

Fill, A. (2001). *Ecolinguistics: State of the art 1998* (pp. 43–53). The ecolinguistics reader: Language, ecology and environment.

Fløttum, K., Dahl, T., & Rivenes, Å. (2016). *Future climate change communication: A linguistic perspective*. Routledge.

Fowler, R. (1991). *Language in the news: Discourse and ideology in the press*. Routledge.

Fowler, R. (2013). *Language in the news: Discourse and ideology in the press*. Routledge.

Garlitos, P. A. L. (2020). Shades of green reporting: A critical discourse analysis of eco-news reports in The Philippines. *Journal of Language and Literature, 20*(1), 25.

Grantham, S., & Vieira, E. T., Jr. (2014). Risk dimensions and political decisions frame environmental communication: A content analysis of seven U.S. newspapers from 1970-2010. *Applied Environmental Education & Communication., 13*, 91–98.

Grundmann, R., & Krishnamurthy, R. (2010). The discourse of climate change: A corpus-based approach. *Critical Approaches to Discourse Analysis across Disciplines, 4*(2), 125–146.

Halliday, M. A. K. (1967). Intonation and grammar in British English. The Hague: Mouton.(Janua Linguarum series Practica 48). *Collected Works, 6*, 1967–1968.

Halliday, M. A. (1978). *Language as social semiotic* (p. 136). Arnold.

Halliday, M. A. K. (1985). Systemic background. In J. D. Benson & W. S. Greaves (Eds.), *Systemic perspectives on discourse* (Vol. 1). Ablex.

Halliday, M. A. K., & Webster, J. J. (2007). Language and education.

Haugen, E. (1972). The ecology of. *Language*.

Heinderyckx, F. (2021). Beyond the news and opinion dichotomy. In S. J. A. Ward (Ed.), *Handbook of global media ethics*. Springer.

Huang, X., & Che, S. (2024). A corpus-based critical discourse analysis of Chinese and American news coverage on climate change. *International Journal of English Linguistics, 14*(3), 42–53.

Ilyinova, E. Y., & Kochetova, L. A. (2024). Mediatization of artificial intelligence concept in the Russian language media discourse: Corpus-based approach. *Vestnik Volgogradskogo gosudarstvennogo universiteta. Seriya 2. Yazykoznanie [Science Journal of Volgograd State University. Linguistics], 23*(5), 108–123. https://doi.org/10.15688/jvolsu2.2024.5.9

Liu, B. (2012). *Sentiment analysis and opinion mining*. Morgan & Claypool.

Marchi, A. (2022). Corpus linguistics in the study of media. In A. O'Keeffe & M. J. McCarthy (Eds.), *The Routledge handbook of corpus linguistics* (2nd ed.). Routledge.

Martin, J. R. and White, P. R. R. (2005). *The Language of Evaluation: Appraisal in English*. Palgrave Macmillan, London.

Moser, S. C. (2010). Communicating climate change: History, challenges, process and future directions. *Wiley Interdisciplinary Reviews: Climate Change, 1*(1): 31–53.

McEnery, T., & Hardie, A. (2012). *Corpus linguistics: Method, theory and practice*. Cambridge University Press.

Milstein, T. (2009). Environmental communication theories. *Encyclopedia of communication theory, 2*, 345–349.

O'Neill, S., Boykoff, M., Niemeyer, S., & Day, S. A. (2015). On the use of imagery for climate change engagement. *Global Environmental Change, 35*, 375–386.

Orhewere, J. A., & Olley, W. O. (2023). The construction of climate change discourse in online news comment sections: A critical discourse analysis. *African Journal of Social Sciences and Humanities Research, 6* (4), 75–84.

Pang, B., & Lee, L. (2008). *Opinion mining and sentiment analysis*. Foundations and trends®. *Information Retrieval, 2*(1–2), 1–135.

Partington, A., Duguid, A., & Taylor, C. (2013). *Patterns and meanings in discourse: Theory and practice in corpus-assisted discourse studies (CADS)*. John.

Pepermans, Y., & Maeseele, P. (2018). Manufacturing consent: Rereading news on four climate summits (2000-2012). *Science Communication, 40*(5), 621–649 (Original work published 2018). https://doi.org/10.1177/1075547018798119

Ramanathan, R., & Hoon, T. B. (2015). Application of critical discourse analysis in media discourse studies. *3L. The Southeast Asian Journal of English Language Studies, 21*(2), 57–68.Benjamins.

Richardson, J. E. (2017). *Analysing newspapers: An approach from critical discourse analysis*. Bloomsbury Publishing.

Sajid, M. A., Anwar, B., & Ashraf, M. (2019). Politics, ideology and print media: A CDA of newspapers' headlines.

Schiffrin, D., Tannen, D., & Hamilton, H. E. (Eds.). (2015). *The handbook of discourse analysis* (2nd ed.). Oxford: Wiley Blackwell.

Schmidt, A., Ivanova, A., & Schäfer, M. S. (2013). Media attention for climate change around the world: A comparative analysis of newspaper coverage in 27 countries. *Global Environmental Change, 23*(5), 1233–1248.

Song, Yunya, Huang, Zeping, Schuldt, Jonathan P and Yuan, Y Connie. (2021). National prisms of a global phenomenon: A comparative study of press coverage of climate change in the US, UK and China. *Journalism*, 1–22. SAGE. https://doi.org/10.1177/1464884921989124

Steffensen, S. V., & Fill, A. (2014). Ecolinguistics: the state of the art and future horizons. *Language sciences, 41*, 6–25.

Stibbe, A. (2015). *Ecolinguistics: Language, ecology and the stories we live by*. Routledge.

Stubbs, M. (1996). *Text and corpus analysis*. Blackwell.

Taboada, M., Brooke, J., Tofiloski, M., Voll, K., & Stede, M. (2011). Lexicon-based methods for sentiment analysis. *Computational Linguistics, 37*(2), 267–307.

Thelwall, M., Buckley, K., & Paltoglou, G. (2010). Sentiment in twitter events. *Journal of the American Society for Information Science and Technology, 62*(2), 406–418.

Wagner, P., & Payne, D. (2017). Trends, frames and discourse networks: Analysing the coverage of climate change in Irish newspapers. *Irish Journal of Sociology, 25*(1), 5–28.

White, P. R. (2008). Evaluative semantics and ideological positioning in journalistic discourse: A new framework for analysis. In *Mediating ideology in text and image* (pp. 37–67). John Benjamins Publishing Company.

Wodak, R. (2001). The discourse-historical approach. *Methods of critical discourse analysis, 1*, 63–94.

Wodak, R. (2009). *The discourse of politics in action: Politics as usual*. Springer.

Wodak, R., & Meyer, M. (Eds.). (2009). *Methods for critical discourse analysis* (2nd ed.). SAGE Publications.

Wodak, R. (2015). Critical Discourse Analysis. In D. Schiffrin, D. Tannen, & H. E. Hamilton (Eds.), *The handbook of discourse analysis* (2nd ed., pp. 504–520). Wiley-Blackwell.

Ye, M., & Friginal, E. (2024). Portrayals of Chinese companies in American and British economic news tweets during China's macroeconomic transitions 2007–2023. *Humanities and Social Sciences Communications, 11*(1), 1–12.

3

Linguistic Strategies in Climate Change Discourse

Introduction

This chapter explores the linguistic strategies embedded within Malaysian climate change opinion news discourse by focusing on the keyword and frequency patterns in two newspapers: Malaysiakini, representing the alternative/tabloid press, and New Straits Times (NST), as a mainstream newspaper. Using corpus-assisted techniques, the analysis uncovers the dominant lexical choices and discursive orientations that characterise the way climate change is presented to Malaysian publics. The contrast between Malaysiakini and NST provides insights into how alternative versus mainstream media negotiate authority, urgency, and responsibility within environmental communication. This chapter starts with describing the corpus information for both newspapers, followed by the keyword and frequency observations.

N. F. M. Nor et al., *A Critical Analysis of Climate Change Opinion News in Malaysian Media*, https://doi.org/10.1007/978-3-032-10333-8_3

Corpus Information

This section explores the basic corpus information of Malaysiakini as the tabloid newspaper, and New Straits Times as the mainstream newspaper. This is crucial to understand the distribution and broadness of the corpus.

A fundamental component of corpus-assisted/based research is basic corpus statistics. Details like the number of files, total tokens/types, and type-token ratio TTR are required for a variety of methodological as well as interpretive reasons.

First, the number of files shows the corpus structure and breadth. In the present analysis, the corpus is split between two outlets namely New Straits Times for mainstream discourse and Malaysiakini for tabloid discourse. Reporting the number of texts from each outlet keeps the dataset balance transparent and helps readers assess sample representativeness. Unchecked unevenness of file count across corpora might skew results, since more texts implies greater lexical variety.

Second, token count (the number of running words in corpus) tells us how large the dataset is and thus, how robust our results are. Generally, larger corpora produce stable frequency results. Token counts in this analysis are used for normalisation and raw frequencies are expressed per thousand words. Without such normalisation, comparing keyword frequencies across NST and Malaysiakini would be misleading as differences could simply reflect differences in corpus size rather than linguistic patterns.

Third, the type count (number of unique word forms), indicates how rich the dataset's vocabulary is. More types signify a larger lexical repertoire which may indicate broader or varied discourse. For example, a higher type count outlet might draw on a broader set of lexical resources for framing climate change, a lower type count outlet might employ more repetitive speech.

Related to this is type-token ratio (TTR), which divides types by tokens to assess lexical diversity. TTR depends on corpus size but is still useful when analysed alongside other statistics. The greater variety of vocabulary usually implies a higher TTR while the lower one indicates more repetition of a term. Similarly in climate change discourse, variation in TTR between mainstream and alternative media may reveal stylistic differences, for instance, mainstream news may repeat official or government terminology while alternative media use a broader lexical set to highlight multiple perspectives.

Additionally, raw frequency refers to the absolute number of times a particular word or expression occurs in the corpus. For instance, if the word climate appears 250 times in a corpus, this figure represents its raw frequency. Raw frequencies are useful in providing an immediate sense of the prominence of a lexical item within a dataset. However, raw counts alone can be misleading when comparing across corpora of unequal size. For example, if the corpus A contains more tokens overall than the corpus B, it is natural that raw frequencies for most words will also be higher, regardless of their actual relative importance in discourse.

To address this limitation, corpus linguists employ normalised frequency, which adjusts raw counts relative to the size of the corpus. Normalisation is usually calculated per 1000 or 1000,000 words (tokens). The formula is

$$\text{Normalised Frequency} = \text{Total Tokens}/\text{Raw Frequency} \times N(\text{standardised based, for instance, } 1,000 \text{ or } 1,000,000)$$

This enables researchers to compare word relative frequencies within a corpus of different lengths. For example, if the word 'climate' occurs 250 times in corpus A out of 500,000 tokens and 400 in corpus B out of 800,000.000 tokens, then normalised frequencies are:

$$\text{Corpus A} : (250 \div 500,000) \times 1,000 = 0.5 \text{ per } 1,000 \text{ words}$$

$$\text{Corpus B} : (400 \div 800,000) \times 1,000 = 0.5 \text{ per } 1,000 \text{ words}$$

Here, the raw frequency is higher in corpus B but normalised frequency shows that both corpora use the word 'climate' at the same rate as corpus size. This distinction is critical for valid cross-corpus comparison. Therefore, normalised frequency provides an appropriate basis for comparing linguistic patterns across corpora as it takes into account differences in dataset size. Without normalisation, such claims about salience of words might be technical artefacts in corpus composition rather than real linguistic tendencies. This explains why the present analysis also takes into account normalised frequency values when comparing between the New Straits Times (mainstream) and Malaysiakini (alternative) corpora.

Corpus statistics are needed for methodological transparency & replicability. Reporting those figures demonstrates that analysis is grounded in methodological data rather than impressionistic observations. It also lets other researchers assess whether the corpora themselves are representative and comparable. Based on Biber (2012) and McEnery & Hardie (2012), reliability of corpus-based findings is not only dependent on the analytical tools used but also on reporting corpus composition clarity.

To recapitulate, corpus information is a technical issue, not a basis for sound corpus linguistics inquiry. By detailing the number of files, tokens, types, TTR, raw, and normalised frequencies, these techniques ensure that future frequency, keyword, and discourse strategy analyses are valid and interpretable within a broader context of climate change communication.

Comparative Analysis of Corpus Statistics between Malaysiakini and New Straits Times

Presentation and interpretation of corpus statistics is an important step in corpus-assisted discourse studies as these figures form the basis for subsequent keyword, frequency, and discourse analyses (Baker, 2006; McEnery & Hardie, 2012). This analysis compares two corpora: The tabloid media perspective is represented by Malaysiakini, and the mainstream by New Straits Times. The statistical information taken into consideration is the number of files, total token, types, TTR, raw, and normalised frequency. The indicators are examined individually to highlight technical considerations as well as discourse implications.

Number of Files

Corpus drawn from Malaysiakini has much more files than NST. This difference presumably means Malaysiakini published more climate change opinion articles in the investigated period. The larger file count also suggests potentially broader authors, perspectives, and thematic coverage. The smaller number of files in NST reflects the more selective nature of opinion publishing in mainstream press where editorial gatekeeping may limit diversity of voices and conformity with institutional priorities. Such discrepancy also illustrates the need to normalise data for size differences, as larger file counts tend to produce higher raw word frequencies (Baker, 2006).

Total Words Count Token Count (Total Words)

Aside from the greater number of files, the Malaysiakini corpus also has a higher token count than NST. The total number of running words, called tokens, indicates how much text is available for analysis. A larger token size for Malaysiakini increases statistical stability of frequency results as patterns drawn from a larger dataset are less likely to be distorted by outliers. The smaller token count for NST, though still sufficient for analysis, suggests a more compact set of opinion texts. This limitation might result in a more confined narrative structure, in the vein of mainstream journalism whose editorial style values institutional and policy-driven perspectives.

Unique Words—Type Count (Unique Words)

The Malaysiakini corpus also has more types /unique words than NST. High type counts signify more varied lexical repertoire. These finding echoes stylistic tendencies in Malaysiakini, which in turn incorporates multiple viewpoints /different framings/critical perspectives on climate issues. The lower type count reflects an increasingly formalised and institutionally consistent vocabulary at NST which seems to favour repeated use of authoritative terminology. This lexical constraint may reinforce clarity and conformity with official policy discourse but may also limit the range of perspectives represented.

Type-Token Ratio (TTR)

Type-token ratio is the ratio of types to tokens that measures lexical diversity. Analysis reveals higher TTR for Malaysiakini than NST, indicating more word choice variation. This suggests Malaysiakini's climate discourse is linguistically diverse, perhaps due to its wide-open platform for contributors and tendency to question or problematise dominant narratives. However, NST's lower TTR reflects recurring terminology and consistent framing versus institutional discourse. Practically, the TTR discrepancy demonstrates how mainstream versus alternative outlets differ in volume and in linguistic resources employed.

Raw Vs. Normalised Frequency

The discrepancies in corpus size also highlight the importance of distinguishing raw from normalised frequencies for analysis. A raw frequency indicates the absolute frequency of a word or phrase whereas normalised frequency adjusts this figure for corpus size (usually per 1000 or 1000,000 tokens). If not normalised, the larger Malaysiakini corpus will artificially exaggerate the apparent importance of lexical items. Normalised frequency allows more precise comparison. In this analysis, the phrase 'climate change' was analysed and it shows that it occurs more often in one than the other corpus. This methodological step is necessary for valid cross-corpus comparisons, so that differences in word frequency are not just corpus composition artefacts but actual discursive tendencies underlying the comparison (Baker, 2006; McEnery & Hardie, 2012).

Climate Change Discourse Implications

Together, these corpus statistics offer a window onto the larger discourse on climate change in Malaysia. Malaysiakini's larger file count, larger token size, greater lexical diversity, and elevated TTR suggest a pluralistic and varied discursive space. That reflects the outlet as an alternative media outlet that spotlights alternative and sometimes opposing voices on the climate debate. In contrast, NST's smaller corpus, lower lexical diversity, and more repetitive institutional vocabulary reinforce mainstream and government-oriented framings of climate change. These findings foresee differences in later keyword and frequency analyses, in which alternative media emphasise contestation and urgency whereas mainstream media emphasise consistency and conformity with official narratives.

Thus, the comparative analysis of corpus statistics between Malaysiakini and NST is a technical one but also a substantive one that reveals how climate change is discussed in Malaysian opinion journalism. File count, token size, type count, TTR, raw and normalised frequency are corpus information that can be used to establish methodological transparency but also be indicators of stylistic and discursive tendencies. These discrepancies between alternative and mainstream outlets suggest that media type influences public discourse on climate change and highlight the need for careful corpus design and interpretation in climate communication research.

Keyword Analysis

This section describes the keyword analysis that was employed to compare between both datasets of Malaysiakini and New Straits Times. It begins with the explanation of both newspapers' stances, followed by a comparative analysis.

Keyword analysis is one of the central tools in corpus-assisted studies as it reveals the lexical features that distinguish one corpus from another (Baker, 2006; Scott, 2010). By identifying words with much higher frequency than expected when compared to a reference corpus keyword analysis identifies the "aboutness" that is the dominant themes and discursive priorities of a dataset (Stubbs, 2010).

The present analysis used the Lancaster-Oslo-Bergen (LOB) corpus for keyword analysis on the Malaysiakini and New Straits Times opinion articles on climate change. This comparative approach identifies salient lexical items as well as reveals how different media construct climate change discourse through linguistic choice. Rather than focusing on institutional terms such as governance and policy for example, alternative media might highlight urgency, accountability, or grassroots.

Keyword analysis is thus intended for two reasons. Firstly to identify the lexical fingerprints for each corpus. Secondly to interpret how these keywords reflect broader ideological cultural and communicative strategies in Malaysian climate change discourse. Moving beyond raw frequency counts to statistically significant lexical prominence patterns provides the foundation for examining how mainstream and tabloid media frame public opinion on climate change

Statistical Measure in Keyword Analysis

In corpus linguistics, different statistical measures are employed to find out which keywords distinguish a target corpus from a reference corpus. As shown in Tables 3.3 and 3.4, the column labelled Type is the lexical item investigated and Rank is the statistical salience position within the keyword list– higher ranks indicate more distinctiveness. Frequency Target and Frequency Reference measures report the raw token counts for each word in the Target and Reference corpora respectively. Raw frequencies are useful for displaying occurrence but misleading when corpus sizes differ, therefore, Normalised Frequency values are calculated (often per million words) for better cross-corpus comparisons Tables 3.1 and 3.2 (McEnery & Hardie, 2012).

To capture distributional spread, Range Target and Range Reference measures indicate the number of texts where a word occurs and Normalised Range expresses this proportion relative to corpus size (Brezina, 2018). These range measures are crucial for avoiding the over-interpretation of words that may be frequent but restricted to only

Table 3.1 Corpus statistics information of Malaysiakini

Item/ Sub-corpus	Malaysia-kini 2015	Malaysia-kini 2016	Malaysia-kini 2017	Malaysia-kini 2018	Malaysia-kini 2019	Malaysia-kini 2020	Malaysia-kini 2021	Total (Overall sub-corpus)
No. of files	1	17	11	11	34	30	31	**135**
Token	1095	15,010	9081	10,005	33,400	31,182	29,746	**129,519**
Type	495	3735	2591	2624	6104	5385	5543	**12,449**
Type-token ratio	0.452	0.249	0.285	0.262	0.183	0.173	0.186	
Raw frequency	1	59	30	20	62	101	82	
Normalised frequency	0.913	3.931	3.304	1.999	1.856	3.239	2.757	

Table 3.2 Corpus statistics information of New Straits Times

Item/ Sub-corpus	New straits times 2015	New straits times 2016	New straits times 2017	New straits times 2018	New straits times 2019	New straits times 2020	New straits times 2021	Total (Overall sub-corpus)
No. of files	11	5	8	10	25	24	27	**110**
Token	8464	3171	6134	7677	16,707	16,329	19,938	**78,420**
Type	2471	1288	1982	2478	3731	3704	4149	**9072**
Type-token ratio	0.292	0.406	0.323	0.323	0.223	0.227	0.208	
Raw frequency	40	15	19	26	75	63	120	
Normalised frequency	4.726	4.730	3.097	3.387	4.489	3.858	6.019	

Table 3.3 Keyword Analysis of Malaysiakini

No	Type	Rank	Frequency target	Frequency reference	Range target	Range reference	Normalised Frequency target	Normalised Frequency reference	Normalised range Target	Normalised range reference	Keyness (Likelihood)	Keyness (Effect)
1	Malaysia	4	732	0	132	0	4929.127	0	0.978	0	3023.743	0.01
2	Climate	7	544	20	134	8	3663.176	19.599	0.993	0.533	2079.111	0.007
3	Water	12	590	438	82	15	3972.93	429.225	0.607	1	1152.723	0.008
4	Change	14	442	267	127	15	2976.331	261.651	0.941	1	958.178	0.006
5	Per cent	15	224	2	60	2	1508.367	1.96	0.444	0.133	902.279	0.003
6	Lynas	17	195	0	13	0	1313.087	0	0.096	0	804.888	0.003
7	Penang	18	187	0	30	0	1259.217	0	0.222	0	771.858	0.003
8	Sustainable	19	186	0	59	0	1252.483	0	0.437	0	767.73	0.003
9	Environmental	20	191	4	70	2	1286.152	3.92	0.519	0.133	750.448	0.003
10	Pm	21	177	0	96	0	1191.879	0	0.711	0	730.572	0.002
11	Global	22	185	4	70	2	1245.749	3.92	0.519	0.133	725.928	0.002
12	Energy	23	266	89	62	12	1791.185	87.217	0.459	0.8	722.4	0.004
13	Forest	24	216	32	48	10	1454.496	31.359	0.356	0.667	709.546	0.003
14	Environment	25	202	27	77	7	1360.224	26.459	0.57	0.467	674.988	0.003
15	Malaysian	26	153	0	67	0	1030.268	0	0.496	0	631.49	0.002
16	Government	27	441	536	107	13	2969.597	525.261	0.793	0.867	621.048	0.006
17	Palm	30	141	8	26	6	949.463	7.84	0.193	0.4	521.769	0.002
18	Waste	31	189	60	44	14	1272.684	58.798	0.326	0.933	521.411	0.003
19	Carbon	32	144	11	49	5	969.664	10.78	0.363	0.333	517.92	0.002
20	Emissions	33	125	0	33	0	841.723	0	0.244	0	515.903	0.002
21	Forests	34	135	8	35	5	909.06	7.84	0.259	0.333	497.676	0.002
22	Oil	37	169	51	36	11	1138.009	49.978	0.267	0.733	473.144	0.002
23	Updated	39	109	0	108	0	733.982	0	0.8	0	449.857	0.001
24	Pollution	40	116	5	45	4	781.118	4.9	0.333	0.267	438.455	0.002

(continued)

Table 3.3 (continued)

No	Type	Rank	Frequency target	Frequency reference	Range target	Range reference	Normalised Frequency target	Normalised Frequency reference	Normalised range Target	Normalised range reference	Keyness (Likelihood)	Keyness (Effect)
25	Development	41	230	203	82	10	1548.769	198.933	0.607	0.667	405.909	0.003
26	Malaysians	43	97	0	39	0	653.177	0	0.289	0	400.324	0.001
27	Resources	45	139	39	56	8	935.995	38.219	0.415	0.533	397.11	0.002
28	Haze	46	106	6	10	5	713.781	5.88	0.074	0.333	392.309	0.001
29	Wastes	49	81	2	12	2	545.436	1.96	0.089	0.133	315.973	0.001
30	Letter	50	170	140	91	13	1144.743	137.195	0.674	0.867	312.833	0.002
31	Economic	51	184	175	61	11	1239.016	171.494	0.452	0.733	309.521	0.002
32	Radioactive	53	88	12	13	1	592.573	11.76	0.096	0.067	293.05	0.001
33	Covid-19	54	71	0	29	0	478.098	0	0.215	0	293.01	0.001
34	Published	55	150	114	135	9	1010.067	111.716	1	0.6	288.998	0.002
35	Projects	56	88	16	26	7	592.573	15.679	0.193	0.467	278.222	0.001
36	State	58	209	277	71	14	1407.36	271.45	0.526	0.933	273.68	0.003
37	Indigenous	59	75	6	17	4	505.034	5.88	0.126	0.267	268.373	0.001
38	Renewable	60	64	0	26	0	430.962	0	0.193	0	264.119	0.001
39	Biodiversity	63	61	0	23	0	410.761	0	0.17	0	251.737	0.001
40	Communities	64	87	24	39	4	585.839	23.519	0.289	0.267	249.666	0.001
41	Land	65	180	223	65	15	1212.08	218.532	0.481	1	249.412	0.002
42	Sarawak	66	60	0	22	0	404.027	0	0.163	0	247.61	0.001
43	Harapan	66	60	0	16	0	404.027	0	0.119	0	247.61	0.001
44	Project	69	92	32	36	9	619.508	31.359	0.267	0.6	246.763	0.001
45	Technology	70	79	17	48	6	531.969	16.659	0.356	0.4	240.993	0.001
46	Countries	72	171	217	56	10	1151.476	212.652	0.415	0.667	232.294	0.002
47	Floods	73	61	3	31	2	410.761	2.94	0.23	0.133	228.334	0.001
48	Impacts	74	55	0	30	0	370.358	0	0.222	0	226.974	0.001
49	Conservation	74	55	0	26	0	370.358	0	0.193	0	226.974	0.001
50	Management	76	102	59	44	13	686.846	57.818	0.326	0.867	225.417	0.001

Table 3.4 Keyword Analysis of New Straits Times

No	Type	Rank	Frequency target	Frequency reference	Range target	Range reference	Normalised Frequency target	Normalised frequency reference	Normalised range target	Normalised range reference	Keyness (Likelihood)	Keyness (Effect)
1	Climate	1	527	20	110	8	6720.224	19.599	1	0.533	2617.142	0.013
2	Malaysia	2	460	0	110	0	5865.851	0	1	0	2431.269	0.012
3	Environmental	3	248	4	74	2	3162.459	3.92	0.673	0.133	1269.656	0.006
4	Change	4	401	267	108	15	5113.491	261.651	0.982	1	1259.517	0.01
5	Environment	5	268	27	84	7	3417.496	26.459	0.764	0.467	1239.281	0.007
6	Global	6	224	4	75	2	2856.414	3.92	0.682	0.133	1143.611	0.006
7	Carbon	7	222	11	59	5	2830.91	10.78	0.536	0.333	1085.707	0.006
8	Sustainable	8	174	0	60	0	2218.822	0	0.545	0	919.063	0.004
9	Energy	9	236	89	54	12	3009.436	87.217	0.491	0.8	878.279	0.006
10	Emissions	10	137	0	45	0	1747.003	0	0.409	0	723.57	0.003
11	Pollution	11	112	5	40	4	1428.207	4.9	0.364	0.267	550.927	0.003
12	Development	12	203	203	66	10	2588.625	198.933	0.6	0.667	539.462	0.005
13	Plastic	13	119	28	16	8	1517.47	27.439	0.145	0.533	489.465	0.003
14	Water	14	235	438	57	15	2996.685	429.225	0.518	1	435.337	0.006
15	Sustainability	15	80	0	36	0	1020.148	0	0.327	0	422.468	0.002
16	Haze	16	81	6	18	5	1032.9	5.88	0.164	0.333	384.972	0.002
17	Waste	17	111	60	30	14	1415.455	58.798	0.273	0.933	373.477	0.003
18	Mangroves	18	70	0	11	0	892.629	0	0.1	0	369.652	0.002
19	Resources	19	100	39	45	8	1275.185	38.219	0.409	0.533	368.885	0.003
20	Natural	20	148	169	58	15	1887.274	165.614	0.527	1	368.605	0.004
21	Green	21	143	155	37	14	1823.514	151.895	0.336	0.933	365.55	0.004
22	Impact	22	82	13	49	9	1045.652	12.74	0.445	0.6	359.109	0.002
23	Biodiversity	23	68	0	25	0	867.126	0	0.227	0	359.089	0.002
24	Economic	25	146	175	65	11	1861.77	171.494	0.591	0.733	354.612	0.004

(continued)

Table 3.4 (continued)

No	Type	Rank	Frequency target	Frequency reference	Range target	Range reference	Normalised Frequency target	Normalised frequency reference	Normalised range target	Normalised range reference	Keyness (Likelihood)	Keyness (Effect)
25	Countries	26	154	217	62	10	1963.785	212.652	0.564	0.667	341.897	0.004
26	Technology	27	79	17	41	6	1007.396	16.659	0.373	0.4	330.049	0.002
27	Forest	28	82	32	30	10	1045.652	31.359	0.273	0.667	302.422	0.002
28	Renewable	29	56	0	25	0	714.104	0	0.227	0	295.712	0.001
29	Health	30	124	172	37	14	1581.229	168.554	0.336	0.933	277.799	0.003
30	Floods	31	55	3	23	2	701.352	2.94	0.209	0.133	267.262	0.001
31	Ecosystem	32	50	0	26	0	637.592	0	0.236	0	264.025	0.001
32	Warming	33	55	5	33	5	701.352	4.9	0.3	0.333	256.75	0.001
33	Wetlands	34	48	0	7	0	612.089	0	0.064	0	253.463	0.001
34	Nations	36	88	76	56	10	1122.163	74.477	0.509	0.667	249.492	0.002
35	Asean	37	45	0	13	0	573.833	0	0.118	0	237.62	0.001
36	Goals	38	62	21	24	3	790.615	20.579	0.218	0.2	236.614	0.002
37	Covid	39	44	0	27	0	561.081	0	0.245	0	232.339	0.001
38	Billion	40	53	9	28	2	675.848	8.82	0.255	0.133	229.837	0.001
39	Reduce	42	69	40	46	9	879.878	39.199	0.418	0.6	226.991	0.002
40	Science	44	96	128	44	9	1224.178	125.435	0.4	0.6	219.97	0.002
41	Mangrove	46	40	0	8	0	510.074	0	0.073	0	211.215	0.001
42	Disasters	47	43	2	19	2	548.33	1.96	0.173	0.133	210.99	0.001
43	Issues	48	65	40	39	8	828.87	39.199	0.355	0.533	209.612	0.002
44	Impacts	50	39	0	27	0	497.322	0	0.245	0	205.934	0.001
45	Earth	51	91	125	36	14	1160.418	122.496	0.327	0.933	204.994	0.002
46	Planet	52	45	6	30	4	573.833	5.88	0.273	0.267	201.562	0.001
47	World	53	189	644	80	15	2410.099	631.097	0.727	1	201.366	0.005
48	Ecosystems	54	38	0	19	0	484.57	0	0.173	0	200.654	0.001
49	Malaysian	55	37	0	27	0	471.818	0	0.245	0	195.373	0.001
50	Policies	56	55	25	33	7	701.352	24.499	0.3	0.467	194.756	0.001

a handful of texts (Gries, 2008). Moreover, Keyness statistics tell us whether a word is over-represented in the target corpus. Keyness (Likelihood) typically uses log-likelihood to determine statistical significance (Rayson & Garside, 2000), while Keyness (Effect) provides a measure of effect size, such as the log ratio, which highlights the substantive magnitude of difference (Hardie, 2014).

Furthermore, these statistical measures establish what lexical items are distinguishable and ensure that differences observed are statistically reliable and substantive. This combination of frequency/distribution and keyness metrics provides the basis for corpus-assisted discourse analysis, allowing the researcher to go beyond impressionistic observations to empirically grounded insights into studied discourse patterns (Baker, 2006).

Stance of Malaysiakini: Keyword Analysis

The Malaysiakini corpus 2015–2021 demonstrates a linguistic profile distinct from that of mainstream news outlets such as the New Straits Times. The most important keywords are "malaysia" (732), "climate" (544), "water" (590), "change" (442), "per cent" (224), "lynas" (195), "penang" (187), "sustainable"(186), "environmental" (191), "global" (185), "energy"(266), "forest" (216), "environment" (202), "government" (441), and "palm" (141). These lexical choices signal a critical/situated discursive orientation. Malaysiakini presents its climate discourse not as abstract or purely scientific but as a phenomenon related to governance, material resources, local disputes, and moral responsibility.

The top keyword list, 'malaysia', shows how Malaysiakini anchors climate debates in national space. Unlike international outlets that report climate change as a global or planetary issues, Malaysiakini always frames the issue in terms of Malaysian identity and accountability. This anchoring is related to governance accountability as reflected in the high frequency of 'government' and the regular reference to political leadership. Climate change is narrated as more than a scientific or environmental problem. Rather, it is a test of government credibility that calls readers to examine policies and leadership promises. The multiple references to national identifiers suggest Malaysiakini employs climate change reporting to probe political authority.

Meanwhile, the Malaysiakini corpus is full of ecological and resource terms. The repetition of water, forest, palm, energy and environment suggests a discursive strategy for encoding climate change in concrete, everyday resources. The newspaper's frequent use of 'water' for example points to Malaysian experiences of floods, droughts and water shortages, all linked to larger climate impacts. The terms forest and palm point to deforestation, palm oil controversy and land-use change, and energy to disputes over fossil fuel dependence and renewable transition. Such keywords make climate change real to readers, linking it to natural resources, ecological struggles and livelihoods in Malaysian society. With such terms, Malaysiakini resists abstraction of climate discourse by making it visceral, immediate, and relatable.

Most prominent are the very specific controversies included in Malaysiakini's keyword profile which are 'lynas' and 'penang'. These lexical items show how the newspaper plans to anchor broader climate conversations with local disputes. By highlighting the rare-earth processing plant Lynas, Malaysiakini brings environmental, industrial, and political debates together under the label of climate responsibility. Although not directly a climate issue, Lynas' discursive presence reflects a larger logic of environmental concern and resistance. Also, "Penang" is used in relation to flooding, reclamation projects and coastal development, making the state a sign of vulnerability and contested progress. Situating global warming within these familiar flashpoints translates climate change issues into local, politically charged encounters in Malaysiakini.

Malaysiakini's discourse also has a moral and global component. Such terms like climate, change, global, environmental, and sustainable keep popping up in the newspaper's references to local and international conversations about climate responsibility. With this hybrid orientation, Malaysiakini can oscillate between local and global. It bases its reporting on national accountability and local controversy but also appeals to global narratives of stewardship, sustainability, and collective action. Thus, Malaysiakini places its readers as Malaysian citizens as well as global citizens who hold leaders accountable while sharing responsibility for planetary care.

Malaysiakini also employs quantification. The keyword 'per cent' refers to the paper's use of numerical framing of progress and critique of performance to measure progress. Citing percentages as indicators of emissions reduction targets, renewable energy quotas, or policy benchmarks gives Malaysiakini technocratic credibility. Numbers support claims of shortfall or delay, reinforcing the regulator role of the newspaper. Its strategy gives it legitimacy as it combines the emotional and moral urgency of alternative journalism with the quantitative authority of evaluation.

Cumulatively, Malaysiakini's discourse can be described as national and accountable, ecological and tangible, controversy-driven, globally moral and numerically critical. It is a non-hegemonic figure in Malaysian climate communication who tells a complicated story of policy and progress about climate change but as a lived, political issue. Rather than legitimising government narratives, Malaysiakini points out governance gaps, amplifies local disputes, and stresses the urgency of ecological crises. Thus, it helps to construct a public space in which climate discourse remains relevant, urgent, and open for debate.

Stance of New Straits Times: Keyword Analysis

The discursive landscape of the New Straits Times corpus is almost the mirror opposite of Malaysiakini. Its keyword analysis puts "climate" (527), "malaysia" (460), "environment" (248), "change" (401), "environment" (268), "global" (224), "carbon"(222), "sustainable"(174), "energy"(236), "emission"(137), "pollution"(112), "development"(203), "plastic" (119), "water" (235), "sustainability"(80), "haze"(81), "mangroves" (70), "biodiversity"(68) and "warming" (55) as central. The pattern here is not controversy or critique but policy legitimacy, technical rationality, and developmental optimism.

Climate and Malaysia are at the centre of NST's discourse about national responsibility in the face of global crisis. Unlike Malaysiakini, where references to Malaysia suggest accountability and scepticism, NST uses Malaysia to legitimise and celebrate. In its climate coverage, Malaysia is mostly portrayed as a responsible actor adhering to international standards and making progress on environmental stewardship. The articulation of climate with Malaysia creates a narrative in which the nation is seen as modern and globally active, and thereby legitimising governmental policies and leadership.

Equally important is the technical-policy register which constitutes the bulk of the corpus of NST. Recurrence of terms like emissions, carbon, energy, and pollution is a discursive strategy of scientisation in which climate change is treated as a quantitative and governable problem. Emissions become entities that can be counted and lowered, carbon is a matter for measurement and management, transition and optimisation of energy are discussed, and pollution is represented as an issue that can be regulated. This technocratic lexicon projects authority and rationality, placing NST in the epistemic space of science and policy rather than grassroots activism or political contestation. It also depoliticises climate discourse, presenting problems as manageable via technical solutions instead of as governance or systemic crisis symptoms.

Along with this technical register, we find a constant emphasis on developmentalism and green growth. Keywords such as "development", "sustainable", and "sustainability"show how NST integrates climate discourse within Malaysia's economic progress narrative. Not seeing environmental action as a constraint on growth, NST sees it as an enabler. Frequent use of hybrid signifying such as "sustainable development"or "green technology" suggests an attempt to reconcile ecological responsibility with economic ambition. In doing so, NST reproduces what is termed the sustainable development discourse that talks about environmental limits without denying growth its supremacy. For NST, sustainability means not sacrifice but opportunity, integrating climate action with national progress.

The NST corpus also reveals ecological symbolism. Words like "mangroves", "biodiversity", "haze", and "warming" show how the newspaper selectively uses environmental imagery to inform its climate narratives. Mangroves & biodiversity are presented as symbols of Malaysian natural heritage exhibited in an attempt to promote conservation in the country.

As a reminder of lived experience with transboundary pollution, haze links local experience with regional dynamics. Warming connects the national story to global scientific discourse indicating Malaysia's awareness of planetary stakes. Unlike Malaysiakini's local flashpoints like Lynas and Penang, these symbols bolster legitimacy and national pride. This puts Malaysia in the context of an environmental success rather than a country of ecological richness that is preserving and adapting.

Overall, NST's discourse is nationally legitimising, technocratic, developmentalist, and symbolically ecological. The paper shows that climate change is a managed process that can be managed via skilled governance, scientific truth, and sustainable development. It does not promote controversy or critique but reassures its readers that Malaysia complies with global norms and seeks balanced progress. Thus, although some public writers critique, NST can be considered as a climate discourse mainstream legitimiser who tames tensions and projects climate change as an issue of institutional competence not contested politics.

Tabloid Vs Mainstream: What Differs in Linguistic Strategies?

Two very different ways of approaching climate change discourse in Malaysia are revealed by comparing the Malaysiakini and the New Straits Times keyword profiles. While the two newspapers use a common vocabulary based on "climate", "change", "environment", and "sustainable", their emphases and discursive functions differ significantly. The result is a media environment where different political parties and linguistic outlets spin different climate change narratives to Malaysians.

For Malaysiakini, accountability and urgency rule. The recurring use of "malaysia", "government", and "per cent" connects climate debates to political accountability, demanding that leaders be measured against policy promises and results. Its discourse also has material resources rooted in them, as evidenced by the repetition of "water", "forest", "palm", and "energy". All these terms situate climate change within the struggles for resource management and ecological vulnerability. Additionally, Malaysiakini incorporates locally relevant controversies such as "Lynas" and "Penang" that localise the global problem of climate change through nationally relevant disputes. This strategy produces a discourse that is rooted, contentious and relevant to the lived experience of its readers. Thus, Malaysiakini frames climate change less as an abstract planetary crisis but as a socio-political problem of governance failures, ecological crises, and localised conflicts.

For New Straits Times, the discourse is more institutional and technocratic. Its frequent use of "emissions", "carbon", "pollution", and "energy"demonstrates an engagement with the scientific or policy registers of climate communications. Climate change is instead presented by NST as a management problem for which technical and institutional solutions suffice. This is supported by the prominence of development, sustainable, and sustainability as words that embed climate action in Malaysia's growth trajectory. Unlike Malaysiakini, which uses sustainability to question or critique policy, NST uses the word to assure readers that economic progress is compatible with ecological responsibility. A selective use of ecological imagery including "mangroves", "biodiversity", and "haze", further reinforces the newspaper's claim of legitimacy and stewardship, portraying Malaysia as a nation rich in natural heritage but responsible in its environmental management.

The divergence also can be interpreted in terms of audience orientation for the two outlets. Malaysiakini's discourse targets readers as critical citizens who will hold leaders accountable, question official narratives, and hold public debate. Use of controversy, localisation and numerical critique reflect its function as an alternative media empowering the public against institutional power. The New Straits Times, for its part, addresses its readers as national stakeholders in a discourse of shared responsibility for Malaysia's development. By emphasising global alignment, national legitimacy, and technical management NST builds its audience as participants in a national progress project.

These differing strategies illustrate two functions of the Malaysian media in climate communication. On one hand, Malaysiakini creates a discursive space where climate change is politicised, contested, and accountable. Meanwhile, in the NST space climate change is depoliticised, rationalised, and incorporated into developmental agendas, although some contexts are found to involve criticism. This dialectic play produces a complex public sphere. Malaysians are exposed to narratives that amplify urgency and critique as well as narratives that promote stability and legitimacy.

That duality holds great significance for climate communication in Malaysia. Malaysiakini may mobilise citizens for more action but risks alienating readers who want stability or who shake at contentious politics. In contrast, NST's discourse may reassure and legitimise, but it risks ignoring the urgency of climate crises and minimising failures of governance. However, the combination of those two discourses offers competing yet complementary ways of making climate change meaningful, where one party critiques, the other reassures.

In the wider context of global climate communication, this contrast is consistent with other countries. Many alternative and independent outlets emphasise urgency, justice, and accountability while mainstream or state-linked outlets favour developmental and technocratic narratives. The Malaysian case, thus, demonstrates that linguistic strategies not only reflect editorial orientation but deeper tensions between urgency and stability, critique and legitimacy, politics and policy.

Conclusion

This chapter investigated linguistic strategies underlying Malaysian climate change discourse using a keyword and frequency analysis of opinions news in Malaysiakini & the New Straits Times respectively. Findings reveal two distinct discursive orientations, rooted in the respective outlets 'editorial character and institution positioning.

Malaysiakini discourse is about accountability, urgency, and localisation. Its repeated references to "malaysia", "government", and "per cent" signal a persistent concern about governance and policy performance, and its use of ecological and resource terms like "water", "forest", "palm", and "energy"highlights climate change in lived experiences. With local controversies like Lynas & Penang incorporated, Malaysiakini turns global climate issues into easily identifiable Malaysian disputes. Its hybrid strategy with mixing global moral appeals with localised critique and emotive urgency with numerical accountability, places the newspaper in opposition to hegemonic media in Malaysia.

The New Straits Times instead offers an institutional, technocratic and developmentalist discourse. Its keywords like "climate", "Malaysia", "emissions", "carbon", "sustainable", "development", "energy", and "pollution", reflect a policy-driven framing of climate change as a regulation, technology, and growth problem. Instead of ignoring controversy or critique, NST frames climate action as an extension of Malaysia's developmental trajectory and international legitimacy. Mangroves, biodiversity, and haze are selective ecological symbols invoked by the newspaper, which have a tangibility that belies their use only as instruments of stewardship and national pride and not of resistance or accountability.

This contrast demonstrates how the media can play both roles in climate communication in Malaysia. Malaysiakini politicises climate discourse by highlighting gaps between policy promises and reality, whereas NST depoliticises it by projecting a national competence versus global norms image. These discourses form a fragmented but plural public sphere, that exposes Malaysians to competing stories that simultaneously reassure and challenge scepticism.

The broader implications are that climate communication is not monumental in Malaysia but deeply mediated through institutional alignments, audience orientations, and linguistic strategies. The alternative newspaper amplifies urgency and critique and calls for citizen engagement whereas the mainstream press legitimises state action and embeds climate change in developmental and technocratic frames. However, both discourses contribute to making climate change meaningful, whether it is as a contested political problem or managed policy domain.

Like the previous chapter showed, keyword and frequency analyses provide more than descriptive statistics. It explains the ideological and discursive bases of climate communication. It reveals the different strategies Malaysiakini used and NST used to communicate about environmental issues. These findings provide the context for a subsequent chapter that moves beyond lexical strategies towards evaluative and affective dimensions of climate discourse, looking at how sentiments and attitudes are articulated in the media and in turn affect wider communications about climate change in Malaysia.

References

Baker, P. (2006). *Using corpora in discourse analysis.* Continuum.

Brezina, V. (2018). *Statistics in corpus linguistics: A practical guide.* Cambridge University Press.

Gries, S. T. (2008). Dispersions and adjusted frequencies in corpora. *International Journal of Corpus Linguistics, 13*(4), 403–437.

Hardie, A. (2014). Log ratio–An informal introduction. *CQPweb tutorials.*

McEnery, T., & Hardie, A. (2012). *Corpus linguistics: Method, theory and practice.* Cambridge University Press.

Rayson, P., & Garside, R. (2000). Comparing corpora using frequency profiling. In *Proceedings of the workshop on Comparing Corpora* (pp. 1–6). ACL.

Scott, M. (2010). Problems in investigating keyness, or clearing the undergrowth and marking out trails. *International Journal of Corpus Linguistics, 15*(3), 241–266.

Stubbs, M. (2010). Three concepts of keywords. In M. Bondi & M. Scott (Eds.), *Keyness in texts* (pp. 21–42). John Benjamins.

4

From Sentiment to Ideology: An Ecolinguistic Analysis of Environmental Discourse

Introduction

How climate change is framed in public discourse goes beyond simple lexical patterns. It requires a closer inspection of emotional, ideological, and ecological dimensions embedded in texts. Thus, this chapter draws together three complementary analysis strands which are sentiment analysis, critical discourse analysis, and ecolinguistics, in order to examine climate change opinion news in Malaysiakini and the New Straits Times. By combining computational and interpretive methods, this chapter explores linguistic strategies employed by different media as well as wider social, political, and ecological implications of their reporting.

The first section reports sentiment analysis results assessing sentiments towards climate change. From this lens, we see if texts in the two outlets frame more positively, negatively, or neutrally and how evaluative language influences public perception. Sentiment analysis identifies emotional orientations within climate discourse, assisting to reveal whether narratives emphasise urgency, hope, scepticism, or disengagement.

In the following part, critical discourse and ecolinguistics analyses are explained. Critical Discourse Analysis (CDA) places the texts within

N. F. M. Nor et al., *A Critical Analysis of Climate Change Opinion News in Malaysian Media*, https://doi.org/10.1007/978-3-032-10333-8_4

broader political, ideological, and social practice structures. The analysis examines how mainstream and alternative media reproduce/resist dominant discourses about climate change, including governance, responsibility, and economic development discourses. CDA examines how climate narratives are constructed through institutional voices in mainstream media and how alternative framings may emerge in other outlets through pluralistic or oppositional perspectives.

Ecolinguistics, or the study of how language constructs human-nature relationships, was also considered as part of the analysis. This analysis examines, using ecolinguistic principles, whether discourse in these newspapers promotes ecologically helpful worldviews such as care, responsibility, and interdependence or perpetuates destructive frames (exploitation, commodification/anthropocentrism). Ecolinguistics goes beyond sentiment and ideology to ask how language reflects or undermines ecological value.

Collectively, these three approaches explore the discursive construction of climate change in Malaysian mainstream and alternative press. The results presented not only represent differences in tonal composition and lexical framing, but also deeper orientations towards ideological and ecological issues within each newspaper. This chapter triangulates sentiment analysis, CDA, and ecolinguistics to show how climate change is mediated linguistically and what this means for public engagement and environmental communication.

Sentiment Polarity of Malaysiakini

Table 4.1 shows sentiment analysis results from Azure Machine Learning (AML) for Malaysiakini climate change opinion news. Findings reflect an overall evaluative orientation of the coverage with a distribution towards negative, neutral, and positive orientations. The sentiment scores generated by AML range from 0 to 1, with scores close to 1 indicating extreme positivity, scores close to 0 indicating extreme negativity, and scores close to 0.5 representing neutrality (Jelen, 2017).

The dataset is dominated by negative sentiment with 122 articles and average polarity of 0.023. This overwhelming share represents 90%

Table 4.1 Sentiment Polarity of Malaysiakini

Sentiment	News article count	Average sentiment polarity score	Overall sentiment polarity percentage	Token	Type
Negative	122	0.023	90%	119,811	10,546
Neutral	1	0.453	1%	295	172
Positive	12	0.950	9%	9413	2600
Total	135	–	100%	–	–

of the sample shows how Malaysiakini frames climate change in critical or problem-focused terms often highlighting challenges, risks, and governance shortcomings.

Only one article has a neutral sentiment—0.453. The neutral category is considered statistically insignificant and omitted further interpretive analysis because of its marginal presence (1%), and because its sentiment lexicon was not according to AntConc's frequency criterion—matched with three occurrences.

Conversely, positive sentiment appears in 12 articles (9%), with an average polarity score of 0.950. Though relatively small in number, these texts tend to be strongly affirmative and often point to optimism, support or constructive developments in broader climate discourse.

To recapitulate, Malaysiakini's sentiment profile is shaped largely by negative frames, with few but notable positive contributions and an overall mixed bag of neutrality. This distribution suggests a discourse which places critique and urgency at the same time as leaving room, although narrower, for accounts of progress and hope.

Sentiment Lexicons of Malaysiakini: Negative

Table 4.2 shows the top ten negative sentiment lexicons from the Malaysiakini sub-corpus of climate change opinion news. They constitute the core of the negative evaluative landscape and signal just how linguistically constructed unfavourable attitudes and criticisms are.

Its most common lexicon is "long", appearing across multiple grammatical categories (adjective, adverb, and proper noun). Its normalised

Table 4.2 Negative Sentiment Lexicon of Malaysiakini

No	Type	POS	Rank	Frequency	Range	Normalised Frequency	Normalised Range	Strength	Sentiment
1	Long	JJ(76); RB(32); NNP(17)	127	125	57	970.527	0.467	Strongsubj	Negative
2	Serious	JJ(38)	410	38	30	295.04	0.246	Strongsubj	Negative
3	Critical	JJ(31)	517	31	20	240.691	0.164	Strongsubj	Negative
4	Disaster	NN(25); NNP(5)	539	30	12	232.926	0.098	Strongsubj	Negative
5	Lack	NN(26); VBP(4)	539	30	22	232.926	0.18	Strongsubj	Negative
6	Concern	NN(29)	563	29	22	225.162	0.18	Strongsubj	Negative
7	Concerned	VBN(10); JJ(17); NNP(2)	563	29	22	225.162	0.18	Strongsubj	Negative
8	Mean	VBP(3); VB(16); JJ(4); NN(1)	689	24	18	186.341	0.148	Strongsubj	Negative
9	Little	JJ(19); RB(1)	832	20	18	155.284	0.148	Strongsubj	Negative
10	Worst	JJS(18); RBS(1)	873	19	16	147.52	0.131	Strongsubj	Negative

frequency is 970.527 with a normalised range of 0.467 at a frequency of 125. The term is strongly subjective (*strongsubj*), reflecting the persistence and extended challenges often highlighted in the coverage.

The adjective "serious" follows as a highly salient lexicon. It occurs 38 times, but its normalised frequency is 295.04 and its range is 0.246, which indicates the gravity it conveys. This describes the seriousness and urgency often attached to the climate change debate in Malaysiakini.

Ranked third is "critical", which is primarily an adjective. It appears 31 times, with normalised frequency 240.691 and range 0.164, supporting evaluative positions highlighting urgent or problematic aspects of climate change. Its designation as strongly subjective indicates its capacity to shape a crisis discourse.

The lexicon "disaster" occurs 30 times as a noun and as a proper noun and has a normalised frequency of 232.926. Its catastrophic associations are representative of how climate change is framed as a looming or unfolding calamity.

Other negative lexicons are "lack", "concern", "concerned", "mean", "little", and "worst". Though they have lower individual frequencies, their combined presence reinforces a critique, insufficiency, and adversity discourse. All are strongly subjective, meaning they consistently exert an evaluative force in expressing discontent or apprehension.

Collectively, these lexicons show how Malaysiakini frames a strongly negative discourse on climate change characterised by persistent, severe, and catastrophic framings. They highlight inadequacies, risks, and threats and thus create a sense of urgency and critique in the newspaper's opinion coverage.

Sentiment Lexicons of Malaysiakini: Positive

Table 4.3 lists the ten most common positive lexicons from the Malaysiakini sub-corpus. Those are terms that form the paper's optimistic register—moments of endorsement, affirmation, and aspiration in a larger climate change debate.

Most prominent is "support", which occurs 153 times as a noun and verb. Its high normalised frequency 1061.696, classified as strongly

Table 4.3 Positive Sentiment Lexicon of Malaysiakini

No	Type	POS	Rank	Frequency	Range	Normalised Frequency	Normalised Range	Strength	Sentiment
1	Support	NN(3); VB(5); VBP(1)	153	9	5	1061.696	0.417	Strongsubj	Positive
2	Better	JJR(6); RBR(1)	191	7	5	825.764	0.417	Strongsubj	Positive
3	Especially	RB(6)	227	6	5	707.798	0.417	Strongsubj	Positive
4	Smart	JJ(2); NNP(4)	227	6	1	707.798	0.083	Strongsubj	Positive
5	Friendly	JJ(5)	289	5	3	589.831	0.25	Strongsubj	Positive
6	Hope	NN(1); VB(1); VBP(3)	289	5	3	589.831	0.25	Strongsubj	Positive
7	Opportunity	NN(5)	289	5	4	589.831	0.333	Strongsubj	Positive
8	Like	VB(3); IN(1)	352	4	3	471.865	0.25	Strongsubj	Positive
9	Worth	JJ(3); NNP(1)	352	4	2	471.865	0.167	Strongsubj	Positive
10	Advocacy	NN(3)	455	3	1	353.899	0.083	Strongsubj	Positive

Table 4.4 Sentiment Polarity of New Straits Times

Sentiment	News article count	Average sentiment polarity score	Overall sentiment polarity percentage	Token	Type
Negative	91	0.038	83%	65,813	8324
Neutral	1	0.530	1%	409	223
Positive	18	0.897	16%	12,198	3071
Total	110		100%		

subjective, highlights the centrality of endorsement and backing in Malaysiakini's climate discourse. After that, "better" occurs 191 times—often in comparative structures—at a normalised frequency = 825.764. That term suggests improvements and preferable comparisons, and reflects how climate initiatives are framed as advances over less efficient alternatives Table 4.4.

Particularly positive aspects of climate action are highlighted with the adverb especially (227 occurrences/normalised frequency 707.398). Similarly, "smart" (227 occurrences) implies innovation and technological advancement—linking climate solutions to intelligence-driven approaches. "Friendly", although less frequent (5 occurrences), carries a strong normalised frequency of 589.831, contributing to a narrative that describes climate change action in approachable, community-oriented terms.

Other notable lexicons include "hope" (289 times), which is associated with aspiration and optimism; Opportunity to point out favourable possibilities; and "like", for preference or positive comparison. Rather less frequently, however, themes of value recognition and active promotion of climate initiatives are introduced with the words worth and advocacy, respectively.

Together these positive lexicons suggest that although Malaysiakini's coverage is dominated by negative framings, optimism is expressed in recurrent language of support, improvement, and opportunity. Hope, advocacy, and technological innovation show how the newspaper alternates critique with occasional stories of progress and possibility.

Sentiment Polarity of New Straits Times

Climate change opinion news sentiment analysis in New Straits Times shows a pattern in evaluative tone distribution. Most articles have a negative sentiment, with 91 pieces (83%) adopting critical or problem-focused framings. These texts average 0.038 sentiment polarity score, which indicates sceptical, cautionary, or negative perspectives within the sub-corpus.

By comparison, 18 articles (16%) have positive sentiment compared with some of the other newspapers we analysed. These articles with an average polarity score of 0.897 project a more favourable and optimistic view of climate change—sometimes highlighting solutions, chances or institutional responses. This more positive sentiment suggests the New Straits Times offers a slightly more rounded evaluative spectrum, suggesting readers perspectives beyond critique.

Neutral sentiment is marginal, however, represented by one article (1%). It is of limited relevance, like the neutral category in the Malaysiakini, as it did not reach AntConc's minimum frequency of three thresholds for sentiment lexicon. Therefore, it was omitted from further interpretation.

Thus, the New Straits Times sentiment profile is dominated by negative discourse with some positive discourse and marginal neutrality. This combination shows the paper tends to put challenges and risks first but also puts optimism and progress at the heart of framing climate change issues.

Sentiment Lexicons of New Straits Times: Negative

Table 4.5 identifies the most frequent adverse sentiment lexicons within the New Straits Times corpus that contribute to a uniquely critical and adverse narrative. The most frequent lexicon is "long" with 46 instances in the forms of adjective, adverb, and proper noun. Its repeated use paints climate change as a persisting problem. In the next section, themes of upheaval and large-scale damage are reinforced by the word "disaster"

(37 instances). Similarly, the word "critical" was used 23 times to indicate urgency and severity—issues were urgent and needed immediate attention.

Thematically, 23 instances of "lack" represent insufficiency and inadequacy, and 21 instances of "degradation" represent deterioration or decline, often associated with environmental or social contexts. "Poverty" appeared 20 times and extends the discourse by connecting climate change with socio-economic struggles and highlighting inequality and vulnerability Table 4.6.

"Concern" has 14 occurrences, indicating general unease and apprehension in the corpus. Such a term reflects both cognitive and emotional responses to climate change—an important component of the newspaper's evaluative framing. The lexicons "worst" (14 instances) and "serious" (13 instances) reinforce extremity and gravity, while "blame" (12 instances) suggests accountability in relation to negative outcomes.

Together, these lexicons show how the New Straits Times constructs climate discourse with adversity as a vocabulary. It is a problem-oriented narrative, starting with endurance (long) and disaster (disaster) then insufficiency/lack, decline/degradation, and vulnerability/poverty. The overwhelming concern amplifies the anxiety discourse, and terms like worst, serious, and blame add responsibility and intensity. Distinguished by strong subjectivity, these lexicons reinforce negative sentiment dominance within the sub-corpus.

Sentiment Lexicons of New Straits Times: Positive

Different lexicons in the New Straits Times climate change coverage provide a basis for an optimistic and positive tone. This is followed by "like", which occurs 17 times in several grammatical contexts (preposition, adjective, verb). Its general usage and frequency suggest it can be used to indicate approval, preference, or favourable comparisons, and thus embed positivity into the discourse.

The adjective "great" comes in second with nine occurrences and indicates approval or admiration and an evaluative tendency to be

Table 4.5 Negative Sentiment Lexicon of the New Straits Times

No	Type	POS	Rank	Frequency	Range	Normalised Frequency	Normalised Range	Strength	Sentiment
1	Long	JJ(33); RB(12); NNP(1)	184	46	29	661.271	0.319	Strongsubj	Negative
2	Disaster	NN(32); NNP(5)	244	37	12	531.892	0.132	Strongsubj	Negative
3	Critical	JJ(23)	407	23	19	330.636	0.209	Strongsubj	Negative
4	Lack	VBP(4); NN(19)	407	23	13	330.636	0.143	Strongsubj	Negative
5	Degradation	NN(21)	452	21	13	301.885	0.143	Strongsubj	Negative
6	Poverty	NN(19); JJ(1)	480	20	9	287.509	0.099	Strongsubj	Negative
7	Concern	NN(12); NNP(1); VB(1)	680	14	13	201.256	0.143	Strongsubj	Negative
8	Worst	JJS(14)	680	14	14	201.256	0.154	Strongsubj	Negative
9	Serious	JJ(13)	728	13	12	186.881	0.132	Strongsubj	Negative
10	Blame	VB(5); NN(6); VBP(1)	786	12	8	172.505	0.088	Strongsubj	Negative

Table 4.6 Positive Sentiment Lexicon of the New Straits Times

No	Type	POS	Rank	Frequency	Range	Normalised Frequency	Normalised Range	Strength	Sentiment
1	Like	IN(13); JJ(1); VB(3)	97	17	11	1487.705	0.611	Strongsubj	Positive
2	Great	JJ(9)	184	9	7	787.608	0.389	Strongsubj	Positive
3	Support	VB(3); NN(6)	184	9	4	787.608	0.222	Strongsubj	Positive
4	Especially	RB(8)	203	8	7	700.096	0.389	Strongsubj	Positive
5	Best	JJS(5); RBS(1)	275	6	4	525.072	0.222	Strongsubj	Positive
6	Well-being	NN(5); VBG(1)	275	6	4	525.072	0.222	Strongsubj	Positive
7	Opportunity	NN(5)	348	5	5	437.56	0.278	Strongsubj	Positive
8	Vital	JJ(5)	348	5	3	437.56	0.167	Strongsubj	Positive
9	Better	RBR(1); JJR(2); RB(1)	442	4	3	350.048	0.167	Strongsubj	Positive
10	Exemplary	JJ(4)	442	4	3	350.048	0.167	Strongsubj	Positive

optimistic. Similarly, the lexicon "support", also appearing nine times, highlights themes of encouragement and endorsement, signalling the paper's tendency to frame certain climate initiatives or actions positively.

The adverb "especially" appears eight times, highlighting noteworthy aspects of the discourse. Its usage brings out specific positives and reinforces the affirmative tone of coverage. Lexicons like "best" and "well-being", which have six instances each, add more positive energy: whereas "best" places excellence or superiority at the centre of attention, "well-being" places emphasis on health, prosperity, and quality of life.

Other important lexicons include "opportunity" (five instances), indicating favourable prospects or possibilities, and "vital" (five instances), indicating the importance and necessity of positive action. The last two are "better" and "exemplary", with four occurrences each. "Better" signifies improvement and progress, whereas "exemplary" signifies admiration and outstanding qualities.

Collectively, these ten lexicons produce a discourse that alternates the dominant negative tone of the newspaper with strong, affirmative passages. Including terms such as support, admiration, well-being, and opportunity the New Straits Times constructs a more balanced and constructive account of climate change that points readers towards challenges but also excellence, progress, and potential for good.

Tabloid Vs Mainstream: What Differs in Sentiments?

Sentiment analysis of Malaysiakini and the New Straits Times on climate change opinion news convergence and divergence. In both newspapers, negative sentiment is evidently predominant—climate change is always presented as serious and problematic. However, how much negativity is foregrounded varies. In Malaysiakini, 90% of articles are negative with an average polarity score of 0.023, a discourse saturated with critique, urgency, and systemic shortcomings. While still strongly negative at 83%, NST offers a relatively broader evaluative spectrum with 16% of its coverage being positive. Optimism when expressed is well articulated—as measured by the mean polarity score of 0.897, and associated with solutions, chances, and institutions. In both outlets, neutral sentiment is tiny at 1% and excluded from interpretation because of statistical significance.

The difference becomes clearer when considering the lexicons that support the two outlets. On the negative side, both newspapers use terms long, serious and critical to indicate persistence, urgency and severity. And in Malaysiakini, they are reinforced by words like disaster, lack, and concern which draw attention to catastrophic framings and to deficiencies in climate governance. And the use of terms like worst and mean adds to the critique: climate change is both a threat and site of failure. NST shares some of these key lexicons but adds degradation and poverty to its negative discourse and explicitly links environmental decline to social & economic vulnerability. Additionally, the lexicon concern within the NST corpus demonstrates the breadth of unease and anxiety underlying its discourse. Terms such as blame further highlight accountability by specifying who is liable for adverse outcomes.

Positively, Malaysiakini lexicons are more sparse but still meaningful. Terminologies like support, better, especially and hope indicate endorsements, improvements, and aspirations. However, such instances appear occasionally and offer isolated glimmers of optimism in an otherwise critical narrative. NST demonstrates more variety and frequency in its positive lexicon. Words including like, great, best, well-being, opportunity, vital, better, and exemplary make up an increasingly plural affirmative register. Together, these terms signify admiration, progress, and resilience and introduce concepts of holistic welfare and institution responsibility. In doing so, NST builds optimism systematically across its climate coverage, recognising risks while pointing to possible solutions and progress.

In this sense, the two newspapers are therefore in different places on the climate change discourse spectrum. Malaysiakini has its roots in critique, a negative sentiment and crisis-oriented lexicons that stress urgency, failure, and risk. Optimism exists but is peripheral. NST is also predominantly negative but tempers its critique with stronger and more diverse positive lexicons that include optimism and progress alongside considerations of risk and vulnerability. This reflects their broader editorial orientations: The other outlets are Malaysiakini as an alternative that champions dissent and critique and NST as a mainstream newspaper that reconciles critical discourse with institutional voices of reassurance and development.

To recapitulate, both outlets make climate change a serious problem, but where Malaysiakini favours urgency and systemic critique, NST balances anxiety with optimism and embeds progress and possibility within its evaluative landscape.

Critical Discourse Analysis and Ecolinguistic

The section that follows goes beyond numerical evaluation of sentiment analysis to provide in-depth analysis of four selected lexicons. Two of these are from *The New Straits Times* (one strong positive sentiment, one strong negative sentiment), and the other two are from *MalaysiaKini* (also one strong positive sentiment and one strong negative sentiment). This selection of two positive lexicons and two negative lexicons aim to capture insights of these lexicons in the discourse context.

Fairclough's three-dimensional model is used to deconstruct each lexicon. This model allows for a systematic examination of the text on three levels:

1. Textual Analysis-identify specific linguistic features such as vocabulary choices, metaphors, and grammatical structures that shape the meaning.
2. Discourse Practice-consider how the lexicons were produced and consumed, examining their journalistic context and potential audience.
3. Sociocultural Practice-situate the language within its broader social and political context, revealing the power relations and ideologies at play in the discourse on climate change.

The critical discourse analysis is extended by using Stibbe's (2015) ecolinguistics framework (eight categories) that will enable readers to have an ecological perspective. The eight categories of Stibbe's framework (chapter 2, Table 2.1) are used to evaluate each lexicon. Description for each category is as follows:

1. Stories we live by: What fundamental beliefs or narratives about the environment are being promoted? (e.g. Human as conquerors of nature, nature as a resource)
2. Framing: How is the environmental issue being framed? Is it an economic problem, a technological challenge, or an ecological crisis?
3. Metaphors: What metaphors are used (e.g., "climate war", "green gold") and what do they reveal about the relationship with the natural world?
4. Ideology: A story of how the world is and should be which is used by members of a group (discourses, i.e. cluster of linguistic features characteristically used by the group)
5. Identities: What identities are constructed for people in relation to the environment (e.g., "consumer", "activist", "expert")?
6. Evaluations: What values are being assigned to environmental actions or entities?
7. Erasure and Salience - What is being left out of the discourse? What is being brought to the forefront?
8. Convictions: What beliefs are being presented as common sense or indisputable truths?

Analysis of Strong Positive Lexicon from the News Straits Times

The lexicon *support* received a high positive sentiment analysis score using Azure Machine Learning (AML) with a reading of strong subjectivity.

The context in which the lexicon *support* is used is shown in text 1.

Text 1 The prime minister has agreed to extend the fund for five years with an additional allocation of RM5 billion. The extension will definitely benefit companies that are producers and users of green technology. This national initiative has opened great business opportunities for the people, and also encouraged them to appreciate the environment and *support* sustainable development in Malaysia.

Textual Analysis

"support" carries a positive connotation and is surrounded by other words that have positive connotations which are 'agreed', 'extend', 'opportunities', 'appreciate', 'sustainable', and 'development'. The Prime Minister who represents the government is the agency that launches the green initiatives, focusing on encouraging companies that practice green technology. Since it is a national initiative, the word *support* also conveys certainty and authority.

Sentence Structure and Coherence

The sentences are declarative and they deliver factual information. The use of declarative sentences and direct statements present factual information.

The sentences are coherent with logical progression, beginning with the Prime Minister's decision about the RM5 billion allocation for green effort, to the beneficiaries (companies and producers of green technology) and finally to the people. Business opportunities for the people implies additional relationships that highlight positive outcomes for the industry players and the society. The overall tone of the text is positive, focusing on positive development of sustainability practices and development in Malaysia.

Discourse Practice

Text and Context

The context of the text is institutional, originating likely from a government body (such as The Ministry of Finance or The Ministry of Science, Environment and Energy), or maybe the Prime Minister's office because the tone is authoritative and focuses on national initiatives. The speaker is the Prime Minister, and this is explicitly mentioned at the beginning of the sentence. The sentence also hints at the possibility of the statement coming directly from the Prime Minister or is a press release/report about his decision. The purpose of the text is to announce and justify the government's decision with the aim of informing the public, giving assurance to the stakeholders (companies, green producers) about the green initiatives. The key purpose is to garner support for the government's green policies. The intended audience is likely a broad audience of readers, which includes the general public, the people, businesses (especially those in the green technology industry), NGOs (environmental NGOs), and the media. This statement is also meant to gain the public and the industry player's trust in the government's green initiatives and policies.

Intertextuality

Since the text is from the government, the text is considered a political genre in the form of press releases or press reports. It resonates language found in other government statements that promote the government's policies, plans and initiatives. In this text, it promotes economic development and environmental protection. The text draws upon 'green discourse' with phrases such as 'green technology' and 'sustainable development' which are prevalent concerns globally and appear in green policy. The text also aligns with the national economic discourse, i.e. business opportunities as seen in the phrase, "*opened great business opportunities for the people*".

Social Practice

Social and Cultural Context

Environmental concern is the focus of the government initiatives that reflect a growing national and global concern of environmental issues and the urgent need for sustainable practices and development. Economic development is also mentioned directly and this can be found in the second sentence—"The extension will definitely benefit companies that are producers and users of green technology" and in the third sentence—"This national initiative has opened great business opportunities for the people".

The Agency (Government)

The text reinforces the government's role as the key driver of both economic development and green initiatives, hence emphasising the government's plan of developing the country's national economy. It also strengthens the government's power because the Prime Minister is portrayed as a benevolent leader.

Public Discourse

The lexicon 'support' which has strong positive sentiment is supported by the lexicon 'appreciate' to indicate the government's move to encourage pro-environmental behaviour among the public.

Ideologies and Power

The text demonstrates the power of the government which is represented by the Prime Minister in the action of allocating a huge amount and significant financial resources for one of the national priorities, i.e., green policy.

The third sentence—"This national initiative has opened great business opportunities for the people, and also encouraged them to appreciate the environment and support sustainable development in Malaysia", emphasises business opportunities that further strengthen the government's power. It suggests an economic ideology by the government that cares about the people while upholding environmental policy nationally. The overwhelmingly positive tone of the text aims to build confidence among the public and support for the government's agenda, which has the potential to dissent critical voices on the government's green plan and policy, and sustainable development in Malaysia. The text is a carefully constructed government discourse designed to announce, justify, and promote a significant national plan. The text reinforces the authority and the legitimacy of the government in leading environmental initiatives in the country.

The analysis continues with the application of Stibbe's ecolinguistics framework (2015), which encompasses eight stories.

Text: The prime minister has agreed to extend the fund for five years with an additional allocation of RM5 billion. The extension will definitely benefit companies that are producers and users of green technology. This national initiative has opened great business opportunities for the people, and also encouraged them to appreciate the environment and support sustainable development in Malaysia.

Stories we Live by

The dominant narrative or story in the text reveals how environmental protection and economic growth mutually reinforce each other. Words like 'extend', 'additional', 'benefit', 'opportunities', 'appreciate', 'support', 'sustainable', are used to frame the 'green economy' and 'green technology' narrative. If the 'green technology' policy and plan proposed by the government is supported, there are 'business opportunities' that can benefit industry players, producers of green technology and the public. This will lead to the achievement of environmental goals (appreciate the environment and support sustainable development). The primary agent is the government, and the beneficiaries are "companies", "the people", and the nation's economy.

Framing

The text is a press release of the Prime Minister's statement, which is considered a piece of government communication and has been carefully designed to announce, justify, and promote a significant national initiative on green development and business opportunities in the green sector. The text is developed in such a way that it begins with the government's announcement to extend the fund with an additional allocation of RM5 million by using words like 'extend', 'fund', 'allocation', 'additional'. The next two sentences that follow elaborate on the business opportunities available with the extension of the green fund allocation—'business opportunities', 'appreciate environment', 'support', 'sustainable development', 'Malaysia'.

Metaphor

The central metaphor in the text is the lexicon 'fund' that can be viewed as a form of sustenance or a lifeline. The 'additional fund' will 'definitely benefit companies'. Money is metaphorically seen as an object that can fuel environmental action, suggesting that financial resources can help solve ecological problems. The phrase "additional allocation" portrays the government as a compassionate entity that provides extra resources. The phrase 'green technology' is metaphorically seen as a business commodity. The environment is treated as a metaphorical resource that can create opportunities to make money, rather than as a living system only. This language subtly reinforces the idea that the environment is valued as primarily economic, and that our relationship with nature is mediated through economy and profit.

Ideology

The lexicon 'support' is surrounded by several words, like 'agreed', 'extend', 'additional', 'benefit', 'opportunities', 'appreciate', and 'sustainable' that have strong positive evaluations. The underlying ideology is that there are plenty of opportunities for business and economic development, which is in line with the government's national agenda, while 'sustainable' is an important word that represents the government's ecological commitment.

Identity

There is a sense of optimism in the press statement that projects the government as a responsible entity and forward-thinking in environmental issues, which are demonstrated in words like 'extend', 'fund', 'additional', 'allocation', 'extension', 'benefit', 'opportunities', 'sustainable', 'development'. The phrase 'national initiative' is salient because the phrase emphasises the government's commitment to the environment and sustainable development in Malaysia. The text aims to foster a sense of shared purpose and patriotism around the green initiative.

Evaluation

The use of words like 'extend', 'benefit', 'opportunities', 'encourage', 'support', 'sustainable', and 'development' denote strong positive evaluations that evoke a sense of progress and possible prosperity with business opportunities for industries and the people. These words also framed the 'green aspect' of the government's environmental effort within a nationalistic framework. There is overwhelming positive evaluation of the initiative which is hoped to lead to a public acceptance of policies.

Erasure and Salience

The press statement erases any likely criticisms that may arise from the government's green policy and initiatives. The statement contains phrases like 'sustainable development', 'business opportunities' that indicate erasure of the government's green policy that might not be economically driven because words that surround the lexicon 'support', like 'extension', 'users', 'green technology', 'national', 'initiative', 'opened', 'great', 'opportunities', 'encouraged', 'support', 'sustainable', 'development', denote encouragement and support from the government to the industries and the public. The specific mention of a number, i.e., 'RM5 billion' is salient and is used to foreground the scale and commitment of the financial investment by the government to support and develop green technology, hence, meant to convince the industries (producers of green technology) and the public.

Convictions

The statement is true with words and a figure that denote positive connotation. Words like 'agreed', 'extend', 'fund', '5 billion', 'benefit', 'opportunities', 'encouraged', 'appreciate', 'support', 'environmental', 'sustainable', 'development', are used to invite and encourage the public to adopt a green habit and environmental appreciation, and support sustainable development. Furthermore, the text aims to instil the conviction among the public that green technology is a positive and economically viable path for the future.

Summary of Analysis

The combined analysis of Fairclough's three-dimensional analysis with Stibbe's ecolinguistics framework demonstrates how ecological "stories we live by" are constructed and reinforced or challenged through language. The linguistic analysis revealed how words are used to evoke positive evaluation in this particular text. The text predominantly promotes ecological ideology whereby environmental action is linked to economic benefits to the industries and the public. The framing of 'green technology' as economic and business opportunities made the initiative more appealing to the audience. The text is a press statement that highlights the Prime Minister as a leader who aligns the national green policy, plan, and initiatives with a global agenda. The text has been carefully constructed as a piece of government communication, to announce, justify, and promote a significant national initiative on green development and business opportunities in the green sector. The text is intertwined with the promotion of the government's national green policy, which is situated within the existing economic paradigm.

Analysis of Strong Negative Lexicon in the News Straits Times

The lexicon *lack* received a high positive sentiment analysis score using Azure Machine Learning (AML) with a reading of strong subjectivity.

The context in which the lexicon *lack* is used is shown in text 2.

Text 2 The number of environmental laws has increased 38-fold globally since the Stockholm declaration on the human environment in 1972. But there has been a *lack* of political will to implement these laws due to the potential impact on livelihoods, lands, properties, and profits. As the first-ever assessment of the global environmental rule of law, the report highlights the vast gaps between legislation and implementation for environmental laws and agencies.

Textual Analysis

The lexicon *lack* is surrounded by words and phrases that imply progress but also hindrance to the progress. The phrase*'38-fold globally...Stockholm declaration on the human environment in 1972'* emphasises a significant and quantifiable authoritative legislative progress. However, the positive development is undermined by the conjunction 'But' in the second sentence which seems to introduce a problem. Words like 'lack', 'gaps', and 'fail', create a negative perception about the effectiveness of environmental regulation, which justifies the strong sentiment of the lexicon 'lack'. The inclusion of 'the global environmental rule of law report' and 'laws and agencies' emphasise authoritative and institutional organisations, rather than individual responsibility.

Sentence Structure

The conjunction 'But' introduces a contrast, suggesting a conflict between legislative effort and action. The phrase 'lack of political will' also identifies the main issue, which is inadequacy in human commitment and implementation of the law. The phrase also presupposes that the failure to implement the law is a known fact. It is linked to the phrase 'Potential impact on livelihoods, lands, properties and profits' to highlight the impact the inadequacy on economy and socio-economic interests.

Discourse Practice

Production

The text appears to be produced by a report or a human rights organisation. It could be a summary of the UN Environment Programme's report, given the formal language and specific references. It aims to highlight the gaps between the law and implementation. It also functions as a critical commentary on global environmental governance. The intended audience is broad, which may include policymakers, environmentalists, researchers, and the general public. The text employs a persuasive narrative, starting with a positive fact about the law and the Stockholm declaration on the human environment. This is followed by a negative reality about the inadequacy of humans in implementing the law through the conjunction 'but' and the phrase 'lack of political will'. It creates a sense of urgency. The tone hints at scepticism because the point made is that while legislative efforts have increased, they are insufficient without effective enforcement and political commitment.

Intertextuality

The text resonates formal language and is likely to be from an authoritative institution, (e.g. UN Environment report). The text draws upon legislative text, with reference to the Stockholm declaration on the human environment in 1972. The text seems to align with the global environment governance.

Sociocultural Practice

Social and Cultural Context

The discourse reveals a fundamental issue which permeates in our lives today, i.e., issues about the environment. The discourse is constructed as a conflict between legislative progress, but the actual implementation is hampered by human inadequacy and political failure or inaction. The text highlights how the "livelihoods, lands, properties and profits" often takes precedence over the implementation of environmental laws. The text also promotes awareness among the general public that policies do not necessarily translate into meaningful change, thereby influencing public perceptions of the government's policy on environmental governance as often ineffective or compromised.

Ideologies and Power

The phrase "lack of political will" is not just a description; it's a political critique, attributing the failure to a deficiency in the leadership's commitment The text demonstrates the power of the government which is represented by the Prime Minister in the action of allocating a huge amount and significant financial resources for one of the national priorities, i.e. green policy. The discourse also reveals a fundamental issue which permeates in our lives today, i.e. issues about the environment. The discourse is constructed as a conflict between legislative progress but the actual implementation is hampered by human inadequacy and political failure or inaction. The text highlights how the "livelihoods, lands, properties and profits" often takes precedence over the implementation of environmental laws.

Stibbe's ecolinguistics framework (2015) analysis is presented in the next section.

Text The number of environmental laws has increased 38-fold globally since the Stockholm declaration on the human environment in 1972. But there has been a *lack* of political will to implement these laws due to the potential impact on livelihoods, lands, properties, and profits**.** As the first-ever assessment of the global environmental rule of law, the report highlights the vast gaps between legislation and implementation for environmental laws and agencies.

Stories we Live by

Generally, the text reveals a significant "story" where the environment is in conflict with the economy. The failure to implement the laws is permitted because of the need to protect human livelihoods and profits, which are presented as a higher priority. The first part of the text that emphasised the increase in the number of environmental laws—"environmental laws has increased 38-fold globally", depicts a narrative of progress and human effort which suggests that humans are actively responding to environmental challenges through legislation. However, the second part, which begins with the conjunction 'but', introduces a story of failure. It implies that despite apparent progress, the real change remains vague. The narrative suggests that environmental protection is an obstacle to socio-economic interests, thus, implying that human economy are primary concerns, while the environment is secondary. The narrative emphasised the gap between legislative intent and action, juxtaposing between 'progress and failure'.

Framing

The issue is framed as a problem of implementation, not of the laws or the broader ecological crisis. The "lack of political will", is framed negatively, focusing on political and human-centred framing. The phrase 'vast gaps' also frames the situation as significant, reflecting failure in the system rather than isolated incidents. This form of framing shifts the focus from ecological destruction onto the failure of the government. However, the progress and increase in laws on environmental protection is framed positively, signalling a sign of awareness and action.

Metaphor

The metaphor lies in the gaps between legislation and implementation, as well as failure in bridging the policy-practice divide. The phrase 'vast gaps' represents the gap between legislation (the progress and promise) and what is delivered (implementation). The phrase "lack of political will", can also be seen as a metaphor that depicts a weak political system or a weak governance, or even a barrier that prevents progress.

Ideology

The discourse contains ideology of environmental protection and progress that has been achieved in environmental laws, but simultaneously criticising the governance/government on prioritising profits and land over ecological sustainability. There is also ideological tension because the discourse that portrays those laws and regulations as necessary, but insufficient without political will and societal change. The government is criticised as prioritising profits over the environment.

Identity

Environmental governance is constructed as a progressive institution on one hand. On the other hand, it is also portrayed as an ineffective actor, due to 'lack of political will'. The text also appears to hold the actors responsible.

Evaluation

The situation is evaluated as serious because the implementation of the environmental laws is not effectively executed due to "lack of political will" and "vast gaps" between legislation and implementation. The text also highlights governance problems and a call for urgent action to overcome the challenges in implementing environmental laws.

Erasure and Salience

The ecological issues that the established laws are meant to address are erased. There is erasure of the political actors or governments that play a significant role in opposing or hindering the environmental laws, subtly constructed in the use of the phrase "lack of political will". The economic interests (profits, lands, properties) are acknowledged at surface level only, thus erasing economic roots behind the political reluctance. The social and cultural factors are also absent, thus simplifying the narrative to a matter of political choice or lack of "political will".

The discourse foregrounds the increase in the numbers of laws (38-fold) and the economic impacts. It draws attention to the large scale of legislative effort and the progress as an achievement. However, salience is shifted in the second part with the mention of 'the lack of political will' and the 'vast gaps'. This shifted the progress to deficiencies and failures in the implementation system.

Convictions

There is a sense of urgency and concern depicted in the phrase "lack of political will" and "vast gaps", that express the idea that a change in the political governance and system is crucial. The intertextuality element is represented in the mention of the Stockholm declaration that suggests a call for environmental obligation and responsibility.

Summary of Analysis

The analysis revealed how language is used to construct the reality of a specific incident about laws and environment. Fairclough's three-dimensional analysis demonstrates that at the textual level, environmental awareness is shown in the progress made with the globally impressive "38-fold" increase in environmental laws. However, the use of the conjunction 'but' immediately directs attention to the failure in implementing the law, the 'vast gaps', and precedence of "livelihoods, lands, properties and profits" over the environment. At the discourse practice level, the text aims to challenge and criticise the belief that legislation is sufficient. The text has an authoritative tone—citing the Stockholm declaration as a form of intertextuality. From a sociocultural perspective, the discourse reveals an ideological conflict where economic priorities take precedence over environmental concerns. The text also critiques the government for its lack of "political will" and political commitment to environmental sustainability.

The application of Stibbe's framework reveals the ecological "stories" in the text. The stories we live by echoes the reality that the environment is in a vulnerable state. The primary story we live by is one where the environment is set against the economy, with the latter being the priority. The environmental issue is framed as a problem of governance and economic trade-offs. The text also embeds erasure and salience. Law and economics are salient in the text, while the actual environmental problems are completely erased. The metaphor "vast gaps" significantly represents the failure of the laws that fail to protect the environment.

Analysis of Strong Positive Sentiment from Malaysiakini Newspaper

Two lexicons from a tabloid newspaper are also included in the CDA and ecolinguistics analysis. The two lexicons are taken from Malaysiakini online newspaper. We will begin with an analysis of strong positive sentiment.

The lexicon *hope* received a high positive sentiment analysis score using Azure Machine Learning (AML) with a reading of strong subjectivity.

Text 3 (Malaysiakini Newspaper) In developing infrastructure projects, nature-based solutions should be explored as they can be *cheaper* as well as provide *multiple benefits* to the rakyat. In the analysis of solution options, we ***hope*** that broader and longer-term *values* are included in the analysis of *costs and benefits***.**

Textual Analysis

The verb 'hope' is a modal verb that expresses a desire or expectation. The verb 'explored' which means to examine or discuss something in detail to understand it better, is supported with the modal verb 'should' to express an advice. This indicates or invites consideration rather than obligation. Words and phrases like 'cheaper', 'multiple benefits', 'values', and 'costs and benefits', denote positive meaning and are used to promote the 'nature-based' solutions, hence emphasising the effectiveness and multiple benefits of "nature-based" solutions. These words also strengthened the "nature-based solutions" infrastructure projects that will benefit both the environment and the rakyat. Furthermore, the phrase 'the rakyat' emphasises inclusivity and social considerations. By focusing on such projects, the approach will lead to a resilient and sustainable Malaysia. The phrase "broader and longer-term values" highlights that the "nature-based solutions" infrastructure projects go beyond financial costs because the project includes ecological, social, and cultural values. Phrases like 'costs and benefits' indicate an economically rational decision.

Sentence Structure

The first sentence focused on the subject 'nature-based solutions', while the modal verb'should' indicates a recommendation, rather than a command. The second sentence supports the first sentence, functioning as a subordinate clause that provides the reason for the recommendation. The values and hope of the agent of the action are presented in the second sentence. The grammatical choice of using the modal verb 'hope' indicates a request, rather than a demand.

Discourse Practice

Production

The language suggests that the text is most likely from the policymakers, or organisations, and is a part of a policy recommendation. The infrastructure projects which are nature-based solutions are seen as preferable and in line with the government's strategic agenda aimed at sustainable development. The third person pronoun 'we' suggests a collective voice (the government or an organisation) and may also indicate a collaborative process rather than a top-down process. The modal verb 'hope' signals a tentative position or stance. This means that the "nature-based solutions" project is a recommendation. Concurrently, the discourse subtly acknowledges any resistance that might arise from parties that tend to favour traditional engineered solutions. The text is constructed with the aim of introducing a new perspective into infrastructure development.

Interpretation

The discourse frames the narrative that nature-based solutions can save costs because they are cheaper and provide 'multiple benefits' to the rakyat. This aligns with economic and ecological values of the government. The phrase "broader and longer-term values" encourages stakeholders to consider holistic values about infrastructure projects that integrate environmental, social, and economic perspectives.

Social Practice

Social and Cultural Context

The discourse aims to demonstrate the conflict between an economic-based ideology with an emerging ideology of sustainable development. The phrase "nature-based solutions" is promoted as desirable and necessary, 'cheaper', and "provide multiple benefits to the rakyat". This implies that it is time to change the traditional engineering projects into ecologically sustainable infrastructure development practices. The discourse aligns with the government's sustainability agenda in Malaysia.

Ideology and Power

The discourse mentions the "multiple benefits to the rakyat" which reflects socially inclusive ideology. The 'rakyat' is directly addressed, thus, the 'rakyat' is symbolically positioned as a significant entity that should be considered in any government's plans, policies, and projects. This also implies the government's democratic approach in ruling the country. A shift towards more sustainable, inclusive, and 'long-term values' should be advocated. Therefore, the text is constructed as an attempt to develop decision-making practice that is inclusive and reflects the government's sustainable development policy in Malaysia.

The analysis continues with the application of Stibbe's (2015) ecolinguistics framework.

In developing infrastructure projects, nature-based solutions should be explored as they can be cheaper as well as provide multiple benefits to the rakyat. In the analysis of solution options, we *hope* that broader and longer-term values are included in the analysis of costs and benefits.

Stories we Live by

The story promotes nature-based solutions which are seen as a desirable approach for infrastructure development. It suggests that nature is a better solution with the ability to serve human and economic needs. The text also frames the environment as a 'cheaper' solution and 'costs savings' that offer "multiple benefits to the rakyat" and has "long-term values". The story suggests 'hope' and challenges the traditional economic view of merely gaining benefits towards an economic development that is sustainable and resilient.

Framing

The text frames the use of "nature-based solutions" as advantageous and a rational economic choice. This is achieved in the phrase "should be explored" that promotes "nature-based solutions" as recommendations or guidelines, but not mandatory. Social inclusivity is emphasised in the word 'the rakyat', and 'multiple benefits' will be achieved with these solutions. The decision is placed within the context of a cost-benefit analysis, whereby the phrase "analysis of costs and benefits", indicates that economic calculation has been done within the ecological or green economic framework.

Metaphors

The word 'hope' is metaphorically conceptualised as a positive emotional aspect associated with future hopes. Words like 'benefits' and 'values' are metaphors that indicate positive outcomes or impacts that can be accrued from 'nature-based solutions'. Additionally, the phrase "analysis of costs and benefits" employs a metaphorical framing which treats costs and benefits as quantifiable elements that align with environmental and economic concerns. The phrase "multiple benefits to the rakyat" treats 'the rakyat' as an entity that can be measured against financial costs.

Ideology

The text promotes economic and sustainability-oriented ideology, and long-term ecological and social values, which are achieved through the phrases "multiple benefits to the rakyat" and "long-term values". These ideologies should be prioritised. The text also subtly challenges the short-term and traditional economic-driven values, advocating for inclusivity (benefit for the rakyat) and importance of prioritising the environment.

Identity

The specific identities in the text are the decision-makers (most probably the government) and the audience. The decision-makers are positioned as forward-thinking and responsible, while 'the rakyat' appears in a passive role as beneficiaries in the decision-making process. The phrase 'we hope' implies collective decision and at the same time constructs the policymakers as progressive and committed to sustainable development. The reference to 'the rakyat' denotes a community-centred and people-oriented governance, emphasising inclusivity and collective well-being.

Evaluations

The word 'cheaper' and the phrase 'multiple benefits' imply positive evaluations of solutions that are cost-effective. The verb 'hope', the modal verb 'should', and the phrase "should be explored" indicate solutions that are positive and worthy of consideration. The 'hope' expressed in achieving 'longer-term values' promotes the aim of achieving sustainability development in Malaysia.

Erasure and Salience

Although the text promotes ecological ideology and environmentally sustainable development, the ecological value of nature-based solutions is erased. The phrase "in developing infrastructure projects" potentially erases the context of any environmental degradation, by focusing only on positive benefits. The text erases ecological and social considerations, suggesting instead that they can be incorporated through the recommended approach. This form of text construction highlights only what the discourse views as important while silencing other, potentially more urgent, perspectives. The words 'cheaper', 'costs and benefits', and 'rakyat' are highly salient and promote economic and inclusivity aspects. The phrase "nature-based solutions" are salient and are presented as prominent and desirable. The 'multiple benefits' and 'long-term values' to the rakyat make these aspects outstanding in the government's decision-making process. The verb 'hope' and the third person pronoun 'we' give weight to the moral values and aspirations to the inclusive long-term values for the 'rakyat'.

Convictions

The text tries to convince the audience that 'nature-based solutions' are both economically beneficial and environmentally sustainable. The longer-term values are desirable and feasible through these solutions. The text reflects the conviction that decisions should be based both on practical benefits for human and financial economic gain. The phrase 'nature-based solutions' is prominent and demonstrates that the decision aligns with sustainable development, which is necessary and beneficial for humans and the environment.

Summary of Analysis

Critical discourse analysis demonstrates how the text is developed to promote an ecological economic ideology by advocating nature-based solutions that are cheaper and provide multiple-benefits to the rakyat with longer-term values. The solutions are more advantageous because they consider social inclusivity and promote a sustainable and resilient ideology. Stibbe's ecolinguistic analysis shows that initiatives and solutions based on nature should be explored and are framed as offering multiple benefits to the communities and the environment. The metaphors of 'hope', 'benefits', and 'values' are embedded with the metaphors of 'costs analysis and benefits'. The discourse promotes social inclusivity (the rakyat/people) and sustainability (nature-based solutions).

The analysis that follows is for the negative lexicon from MalaysiaKini, which is *worst*. The lexicon *worst* received a high negative sentiment analysis score using Azure Machine Learning (AML) with a reading of strong negative subjectivity.

The context for the lexicon worst is shown in text.

Text 4 (Malaysiakini Newspaper) The Sibu water board has announced that all the water treatment plants in Sibu have ceased to operate at the moment, and the clean water supply to the whole of Sibu will be affected. This incident shows the *worst* consequence of unabated logging and burning, which is one of the causes of global warming.

Textual Analysis

The word 'worst' emphasises the severity of a condition or situation. It denotes a negative evaluation, from a serious situation to the most serious or extreme situation. The phrase 'shows that' which precedes the word 'worst' is purposely used to demonstrate a cause and effect link. The word 'consequence' indicates causality, while the word 'unbated' implies intensity of the 'logging and burning' activities. The two activities heightened and worsened global warming conditions. Other words like 'ceased' and 'affected' reinforced the negative sentiment analysis of the word 'worst' because both words indicated disruption of the water treatment plant and unfavourable outcome pertaining to the water supply in Sibu.

Sentence Structure and Coherence

The first sentence is in present perfect tense, indicating a recently completed action and its relation to the present. The phrase 'have ceased' showed a recently completed action, while the adverbial phrase 'at the moment' indicated the current time. The first sentence also includes the passive voice structure in the form of future tense as shown in the phrase "will be affected". The second sentence is in simple present tense, demonstrating that the incident has an ongoing or current significance. The first sentence sounds formal; "The Sibu water board has announced that…", establishes the source of information. The first sentence also informed the audience of the impact of temporary work closure of the water treatment plant. The second sentence shifts from the local situation to the environmental context, which suggests that the local incident represents something serious. The second sentence links local environmental damage (logging, burning, water disruption) to global concern (climate change/climate crisis).

Discourse Practice

Producer and Audience

The first sentence is authoritative and formal; "The Sibu water board…". The statement aims at highlighting the severity of the incident. The intended audience are people living in Sibu and are meant to get the attention of the people. The phrases "ceased to operate" and "affected clean water" supply implied unwanted consequences of the ceased operation of the water treatment plant. The second sentence is intended to raise the audience's awareness about global warming because the incident is linked to the consequence of unrelentless logging and burning activities. The use of the word 'worst' invokes emotional engagement and concern about the environment. It also links the infrastructure failure (ceased operation of water treatment plant and water supply disruption) with environmental problems, hence creating an interconnected crisis.

Intertextuality and Discursive Practice

Several genres can be identified in the text with a mix of report (announcement by the Sibu water board), environmental problems (unabated logging and burning), and climate discourse (water disruption is linked to environmental concerns/issue, specifically global warming). The word 'worst' is superlative and is purposely used to highlight and link the severity of the local incident to the global crisis. The discursivity element is demonstrated in the use of the word 'worst' which invokes emotional engagement and concern about the environment.

Sociocultural Practice

Social and Cultural Context

The discourse can be placed within the context of the climate crisis in Malaysia and Southeast Asia (logging, deforestation, open burning, haze, and others). The incident is framed as 'the worst' consequence that implies irrevocable damage to the environment. The word 'worst' also communicates risk and vulnerability. Consequently, the discourse can help shape cultural understandings of environmental damage or risk that may lead to a behavioural change about the environment among humans. The narrative also emphasises the urgency of protecting the environment and a call for more enforcement actions from the authority.

Ideology and Power

The first sentence is formal and authoritative. It emphasises the ideological function whereby the "ceased operation of all the water treatment plants in Sibu" is unavoidable and necessary. However, the second sentence framed the outcome of the incident as intolerable and is linked to destructive and unacceptable environmental practices (unabated logging and burning). This reinforces the ideology that human destruction of the environment can lead to catastrophic consequences. The word 'worst' is used to create emotional engagement, particularly the feeling of fear about global warming which is created by harmful activities like logging and burning. There is a hint at a call for stricter regulation, policies, and enforcement on logging activities in Sibu.

The analysis continues with the application of Stibbe's ecolinguistics framework (2015).

The Sibu water board has announced that all the water treatment plants in Sibu have ceased to operate at the moment, and the clean water supply to the whole of Sibu will be affected. This incident shows the *worst* consequence of unabated logging and burning, which is one of the causes of global warming.

Stories we Live by

The text tells the story of cause and effect, i.e. destructive human activities (unabated logging and burning) that lead to water disruption and worsening global warming. The mention of 'worst consequence' suggests a narrative that implies harmful human activities can eventually result in disastrous consequences on the well-being of humans.

Framing

The incident is framed not only as a human crisis, i.e. water supply disruptions, but also the phrase "worst consequence" that is linked with "unabated logging and burning" suggests environmental crises as severe consequences. The phrase 'unbated logging and burning' also frames these activities as destructive, unacceptable, and has a direct impact on human life and well-being. A global warming frame (worst consequence of unabated logging and burning, which is one of the causes of global warming) is introduced at the end of the text.

Metaphors

The word 'worst' implies a scale from bad to worst. The word 'worst' is metaphorically conceptualised and framed as environmental damage, that has devastating impact on human well-being. The lexicon 'worst' also evokes urgency, specifically pointing to the authority to enforce stricter logging rules and open burning. The 'consequence' of "unabated logging and burning" personifies these processes as unrelenting forces that create a metaphor of unstoppable destruction.

Ideology

The text promotes the ideology of environmental concerns and protection over exploitation of economy. Although the text is identified as a report released by the Sibu water board, there is absence of explicit social actors which means that the real actor is not disclosed. This absence implies that no one is directly blamed for the ceased operations of the water treatment plants and water disruptions. The word 'worst' and surrounding words and phrases like 'affected', 'consequence', 'unabated logging and burning', and 'global warming' reject industrial and economic ideology, by promoting an ecological ideology that views human destruction of the environment as serious and must be urgently addressed. The phrase "unabated logging and burning" also hints at condemnation of human greed for profit over protection of the environment.

Identity

The evident social actor identity in the text is the people of Sibu, who are portrayed as victims of environmental negligence. There is a shared identity of defencelessness among the Sibu community. Their well-being is affected because they are deprived of clean water. The link between the water disruption incident and the phrase "worst consequence" demonstrates the authority's failure in protecting and taking care of people of Sibu's well-being. The "unabated logging and burning" activities are constructed as a damaging force, without revealing the identity of the social actor involved in these environmentally destructive activities.

Evaluations

The word 'worst' denotes a strong negative evaluation that frames the environmental impact as extreme and unacceptable. The word 'worst' has a value judgement that indicates the seriousness of the incident. The word 'unbated' further intensifies the negative evaluation by suggesting irresponsibility and lack of enforcement on logging activities and open burning. In addition, the phrase "the worst consequence" shows how human activities are highly damaging and lead to impact on human well-being (water disruptions) and climate crisis (global warming). The producer of the text condemns logging and burning activities as unsustainable practices that have calamitous outcomes.

Erasure and Salience

Although the text is an announcement made by the authority, i.e. the Sibu water board, the responsible agent is erased. In addition, there is also no mention of the perpetrators who have recklessly destroyed the environment through 'unabated logging and burning' activities. The language used focuses specifically on the negative impacts by erasing possible mitigating factors, such as efforts for reforestation or sustainable logging. The phrase "worst consequence" indicates disaster which ends in global warming. The first sentence focused on the water crisis which is made salient through phrases like 'ceased to operate' and 'clean water supply…affected'. The link to 'worst consequence', highlighted the negative impact of this incident on the well-being of the people of Sibu. The word 'worst' is a form of superlative discourse marker which is used to draw attention to the severity of the event, and the phrase 'worst consequence' amplifies the urgency of reckless logging and burning activities that end in increased global warming.

Convictions

The text conveys a framing conviction to the audience that irresponsible and uncontrolled logging activities lead to degradation of the environment, hence, resulting in global warming. The text also carries the conviction that these activities have impacted the well-being of the people, evidenced by water treatment plants that will be temporarily halted and water disruption to the people of Sibu. In addition, there is an implied conviction that urgent action must be taken to curb these destructive environmental activities.

Summary of Analysis

A critical discourse analysis of Fairclough's three-dimensional framework demonstrates how environmental issues are constructed through language. At the textual level, the word *"worst"* conveys a strong negative evaluation, framing the environmental impact as contributing to global warming, hence, highlighting the impact as a crisis. The surrounding phrases such as "ceased to operate" and "clean water supply… will be affected" emphasise urgency and crisis. From a discourse practice perspective, the text combines institutional reporting (Sibu water board) with evaluative commentary in the second sentence, whereby the water disruption is attributed to uncontrolled logging and burning activities. The discourse also reflects wider environmental concerns in Malaysia, resonating with global narratives of climate crisis and global warming.

The application of Stibbe's ecolinguistics framework demonstrates that the word 'worst' evokes a strong negative connotation and functions as a superlative intensifier that frames the text as an ecological disaster. The water disruption incident is framed as a consequence of uncontrolled deforestation (logging and burning), which has an impact on the people of Sibu. The text also presents the people of Sibu as victims due to irresponsible activities by social actors which are not mentioned (erased category). The text embodies a clear ecological ideology where human and environment are competing over economic exploitation.

Conclusion

In general, sentiment analysis of Malaysiakini and the New Straits Times (NST) reveals the negative framings that dominate Malaysian climate change discourse and the different editorial orientations that shape how negativity and optimism are dispersed. Whereas Malaysiakini projects critique, urgency, and systemic failure as dominant discourse features, NST frames problems with stronger affirmation, opportunity, and institutional reassurance. These findings suggest how sentiment shapes climate narratives, and how evaluative discourse shapes public perceptions of climate change as either a crisis or a challenge with potential for progress. Importantly, polarity and analysed lexicons are but the first layer of meanings. How climate discourse is positioned ideologically and ecologically is dependent on understanding how power, ideology, and worldviews are shaped. The following sections thus turn to critical discourse analysis/ecolinguistics to explore how these texts reproduce/resist dominant narratives and construct human-nature relations that reinforce/challenge ecological values.

The combined application of Fairclough's three-dimensional framework and Stibbe's ecolinguistics framework provides a comprehensive understanding of how attitudes towards climate change are constructed and communicated in opinion news. The analysis demonstrates that micro-analysis of opinion news reveals embedded attitudes and ideologies. Fairclough's three-dimensional model highlights how specific words, phrases, and discourse markers reveal ideology and power conflict, for instance, use of the discourse marker 'but' signals the move from legislative success (environmental laws has increased 38-fold globally, Stockholm declaration) to implementation failure. The implementation failure also unveils the conflict between economic priorities over environmental issues and concerns. Positive sentiment words like 'support' and 'hope' are used to justify and promote the authority's initiatives and plans on sustainable practices. The two texts are intertwined with the promotion of the government's national green policy within the green economic paradigm. The word 'fund' in text 1 is a metaphor that fuels positive actions towards the environment, while words like 'benefits' and 'values' metaphorically signify positive outcomes. The positive 'stories' enlighten the readers on efforts taken to promote sustainable practices, green technology, and environmental protection. The negative 'stories' frame the incident as an ecological disaster.

On the other hand, negative sentiment words such as 'lack' and 'worst', construct a sense of crisis which results in unwanted consequences of events that have an impact on human well-being, like the water disruption in Sibu. The consequences are linked to broader environmental destruction, specifically uncontrolled logging and open burning of forests. Stibbe's ecolinguistics framework elucidates the framing of ecological "stories" that shape our relationship with the environment. The category of salience and erasure is significant because the analysis emphasises how the people of Sibu are victims, while the social actors who are responsible for the 'unabated logging and burning' are not mentioned at all.

The integrated use of these two frameworks demonstrates that opinion news is a complex discourse that frames environmental issues as a consequence of human greed and ideology, where economic benefits and priorities have destructive consequences for both the environment and human communities.

Reference

Fairclough, N. (1995). *Critical Discourse Analysis: The Critical Study of Language*. Longman

Stibbe, A. (2015). *Ecolinguistics: Language, ecology and the stories we live by*. Routledge.

5

An Appraisal Theory Analysis of Opinion News Discourse

Introduction

Understanding how climate change is evaluated in public discourse goes beyond mere sentiment. It also requires systematic exploration of how language reflects attitudes, judgements, and values. Based on findings from the previous chapter on sentiment, this chapter applies Appraisal Theory (Martin & White, 2005) to analyse climate change opinion news in Malaysiakini and the New Straits Times. Although the sentiment analysis revealed general tonal orientations (positive, negative, neutral), appraisal analysis permits a closer examination of how these evaluations are constructed linguistically, how they are dispersed across affect (emotions), judgement (social evaluation of behaviour), and appreciation (aesthetic and value-oriented assessment). This tells us about ideological positioning in the media.

This chapter, using appraisal analysis that focuses on attitude subsystems, examines how the two newspapers frame climate change as a matter of urgency, governance, morality, or opportunity. It examines whether negative or positive attitudes exist and how authors encode dissatisfaction, insecurity, or disapproval alongside support, resilience,

N. F. M. Nor et al., *A Critical Analysis of Climate Change Opinion News in Malaysian Media*, https://doi.org/10.1007/978-3-032-10333-8_5

or endorsement. These are important insights into how opinion news convinces, critiques, or reassures readers.

The chapter centres on an appraisal of resource descriptions from Malaysiakini and New Straits Times, which highlight convergence and divergence patterns between alternative and mainstream media. Continuative comparative discussion links these findings to larger questions of media discourse, ideology, and climate communication in Malaysia. In summation, the analysis demonstrates how appraisal theory might illuminate the complex interaction of language, evaluation, and power in public understandings of the climate crisis.

The first section of this chapter discusses the appraisal analysis for Malaysiakini. Then, it is followed by the appraisal analysis for New Straits Times. Each section starts with the appraisal categorisation (negative, then positive) and is followed accordingly by the excerpts' examples (negative, then positive), by both newspapers.

Appraisal Categorisation: Negative Sentiments and Excerpts in Malaysiakini

See Table 5.1.

Appreciation of negative sentiments in Malay climate change opinion news is distributed across the three core subsystems of affect, judgement, and appreciation that reveal different dimensions of how negativity is constructed.

In the affect subsystem, which codes for emotional responses, the most common categories are dissatisfaction (15 occurrences) and insecurity (nine occurrences), followed by smaller but significant amounts of unhappiness (four occurrences) and disinclination (two occurences). Results indicate that Malaysiakini's discourse frequently reflects discontent and unease—frustration with climate governance and uncertainty about environmental futures.

The judgement subsystem records evaluations of behaviour in relation to social norms and ethics. Normality, capacity, and tenacity appear modestly under social esteem with seven occurrences each indicating assessments of competence and determination in climate action.

Table 5.1 Negative sentiments categorisation in Malaysiakini

Subsystem	Sub-branches	Level 1 negative frequency	Level 2 negative frequency	Total negative frequency
Affect	(a) Happiness/Unhappiness	4		30
	(b) Security/Insecurity	9		
	(c) Satisfaction/Dissatisfaction	15		
	(d) Inclination/Disinclination	2		
Judgement	*Social esteem* (a) Normality (how special/fate/ usuality)	1	7	21
	Social esteem (b) Capacity (how capable/ ability)	3		
	Social esteem (c) Tenacity (how resolute/ resolve/ inclination)	3		
	Social sanction (d) Veracity (how honest, truth/ probability)	1	14	
	Social sanction (e) Propriety	13		
Appreciation	(a) Reaction (impacts and qualities)	8		8
	(b) Composition (balance and complexity)			
	(c) Valuations			
Total of appraisal occurrences				59

The social sanction category is particularly striking, as it appears 13 times (propriety) and veracity once. This strong emphasis on propriety demonstrates how Malaysiakini tends to frame climate change as an ethical question of morality, fairness, and responsibility.

In the appreciation subsystem, processes and phenomena are observed eight times by the reaction category—this shows that the newspaper often points to perceived impacts or quality of climate issues. Composition, balance, and complexity are, however, absent, and this suggests that although Malaysiakini critiques impacts and consequences, it pays less attention to systemic balance or broader assessments of value.

Collectively, these findings indicate that Malaysiakini has the most negative appraisals within the affect subsystem, with dissatisfaction as the strongest evaluative position. This is followed closely by the judgement subsystem—particularly propriety—reflecting ethical critique. In combining emotional dissatisfaction, moral evaluations, and attention to impacts, the newspaper constructs a discourse that frames climate change as a site of crisis as well as a matter of moral responsibility and social accountability.

The first example is the negative sentiment lexicon (NSL) "long", which is shown in Table 5.2. For the affect subsystem, this expression corresponds to dissatisfaction with regard to possible results from misguided investments. In the judgement subsystem, it is considered propriety—a criticism of unethical or irresponsible financial decision-making. This sentence suggests a negative evaluative attitude towards the wasteful allocation of scarce public resources.

In the second example, the NSL is "serious". Similarly, dissatisfaction is grounded in the affect subsystem, this time over disappointment with a missed national opportunity. Tenacity from the Judgment subsystem identifies a perceived lack of commitment to holding relevant actors accountable. This tonal quality points out both the weight of the issue and the disappointment at the lack of follow-through.

Finally, the third example is the NSL "disaster"—that is, the chemical dumping incident at Sungai Kim Kim in Johor explicitly made out to be a manmade crisis. This causes unhappy feelings in the affect subsystem, while the judgement subsystem once again points to propriety and the

Table 5.2 Examples of Malaysiakini's negative excerpts

No	Negative sentiment lexicon (NSL)	context	Affect (emotions felt)	Judgement (of people's behaviour)	Appreciation (of things)	Description
1	Long	For otherwise, we will be making wrong investments *[-judgement: propriety]* that will not be sustainable in the long run, and which would be a colossal waste of scarce public resources	Disinclination	Propriety		[-Affect: Disinclination]—The sentence expresses reluctance or aversion about making wrong investments and the potential waste of public resources [-Judgement: Propriety]—The sentence expresses judgements related to non-ethical behaviour, as it talks about making "wrong investments"
2	Serious	However, the scrapping of the MTHA is a missed opportunity *[-judgement: tenacity]* for Malaysia to show that it is serious about holding its own entities accountable, and that neighbouring countries should do the same	Dissatisfaction	Tenacity		[-Affect: Dissatisfaction]—The sentence expresses disappointment or missed opportunity regarding the scrapping of the MTHA [-Judgement: Tenacity]—The sentence suggests that the scrapping is a missed opportunity and that it is a wrong commitment

(continued)

Table 5.2 (continued)

No	Negative sentiment lexicon (NSL)	context	Affect (emotions felt)	Judgement (of people's behaviour)	Appreciation (of things)	Description
3	Disaster	For instance, the present chemical dumping at Sungai Kim Kim in Johor has been a manmade disaster and an emergency *[-judgement: propriety]*, because a large number of the population faced severe health risks *[-appreciation: reaction]*	Unhappiness	Propriety	Reaction	[Affect: Unhappiness] - The sentence describes a sense of disheartenment regarding the chemical dumping as a manmade disaster, highlighting the severe health risks faced by a large population [Judgement: Propriety] -The sentence addresses the unethical behaviour of chemical dumping [Appreciation: Reaction] - The sentence addresses the severe health risks that chemical dumping has impacted

unethical act. Also activated is the appreciation subsystem through reaction (impact), which points out the serious health risks facing affected communities. This produces a strongly negative appraisal that includes emotional distress, ethical condemnation, and recognition of disastrous results.

Appraisal Categorisation: Positive Sentiments and Excerpts in Malaysiakini

See Table 5.3.

Within the affect subsystem, which captures emotional responses, a total of 26 instances are recorded. These are distributed across security (seven), satisfaction (four), and inclination (15). The predominance of inclination signals strong expressions of desire, endorsement, or willingness in relation to climate action, suggesting that positive attitudes are often framed in terms of forward-looking aspirations and commitments.

The judgement subsystem, which evaluates behaviour against social norms, registers 22 occurrences in total. Under social esteem, the categories of capacity (nine) and tenacity (10) dominate, reflecting positive assessments of competence and perseverance in addressing climate challenges. The social sanction category includes propriety (three), indicating that certain actors are appraised favourably in terms of moral appropriateness and ethical responsibility. Together, these patterns suggest that Malaysiakini frequently frames individuals or institutions as both capable and committed in their climate-related roles.

The appreciation subsystem is marked by 13 instances of valuation, pointing to recognition of the importance and significance of climate-related measures. Interestingly, reaction and composition do not appear in the dataset, showing that less attention is given to the immediate emotional impact of climate actions or to their balance and structural organisation. Instead, evaluations tend to emphasise the broader worth and value of climate initiatives.

In total, 61 instances of positive appraisal were identified. Although negative sentiments dominate the overall corpus, this finding highlights that Malaysiakini also provides space for optimism, encouragement,

Table 5.3 Positive sentiments categorisation in Malaysiakini

Subsystem	Sub-branches	Level 1 positive frequency	Level 2 positive frequency	Total positive frequency
Affect	(a) Happiness/Unhappiness		26	26
	(b) Security/Insecurity	7		
	(c) Satisfaction/Dissatisfaction	4		
	(d) Inclination/Disinclination	15		
Judgement	*Social esteem* (a) Normality (how special/ fate/ usuality)		19	22
	Social esteem (b) Capacity (how capable/ ability)	9		
	Social esteem (c) Tenacity (how resolute/resolve/ inclination)	10		
	Social sanction (d) Veracity (how honest, truth/ probability)		3	
	Social sanction (e) Propriety	3		
Appreciation	(a) Reaction (impacts & qualities)		13	13
	(b) Composition (balance & complexity)			
	(c) Valuations	13		
Total of appraisal occurrences				61

and affirmation. These appraisals capture emotions of hope and joy, acknowledge competence and determination, and foreground the value of specific climate initiatives. Taken together, they demonstrate that positive discourse—though less frequent—plays a meaningful role in shaping the newspaper's climate change coverage. The excerpts' examples are represented below.

Table 5.4 provides several examples of categorisation examination of Malaysiakini's Positive Sentiment Lexicons (PSLs), emphasising their evaluative and affective aspects as components of Appraisal Theory's attitude subsystem. It examines the utilisation of certain positive terms—such as "hope", "opportunity", and "like"—in public discourse to express thoughts, assess behaviours, and acknowledge the importance of objects or events, especially with global or societal issues.

Each PSL is analysed within its particular context, emphasising the linguistic and situational setting in which it occurs. The term "hope" pertains to dialogues regarding technological breakthroughs and human ingenuity aimed at tackling global issues, whereas "opportunity" is connected to political actions in response to reports and recommendations for sustainable transitions. Simultaneously, "like" serves to engage policymakers and civil society leaders in the advancement of sustainable development objectives.

The table classifies the emotional and evaluative functions of these lexicons into three dimensions: affect, judgement, and appreciation. Affect denotes the emotional disposition associated with a term, encompassing positive aspirations or optimism (e.g., "hope" and "opportunity" signify a forward-looking perspective). Judgement assesses individuals' behaviours or abilities, including the government's obligation or capacity to respond effectively to global concerns, as exemplified by "opportunity". Appreciation evaluates the value or relevance of concepts, things, or phenomena, including the role of technology or sustainable economies in tackling global challenges.

The description column offers a comprehensive analysis of the operational mechanisms of each PSL inside speech. "Hope" signifies a favourable disposition towards addressing global challenges, assesses the

Table 5.4 Examples of Malaysiakini's positive excerpts

No	Positive sentiment lexicon (PSL)	Context	Affect (emotions felt)	Judgement (of people's behaviour)	Appreciation (of things)	Description
1	Hope	In current contexts and advancement of technologies and human creativity [+appreciation: valuation], these are actually some of the best means available at our disposal [+judgement: capacity] in the hope [+affect: inclination] of positively solving our global predicaments	Inclination	Capacity	Valuation	[+Affect: Inclination]: Reflects a positive desire or inclination towards achieving a solution to global challenges [+Judgement: Capacity]: Expresses confidence in the capability and effectiveness of the available means [+Appreciation: Valuation]: Evaluates the value and importance of technology and creativity in solving global issues
2	Opportunity	As the first to follow these landmark reports and accompanying recommendations [+judgement: capacity, propriety], this budget is an opportunity for the government to respond to these challenges [+affect: inclination] and start the crucial transition towards a sustainable economy. [+appreciation: valuation]	Inclination	Capacity Propriety	Valuation	[+Affect: Inclination]—Expresses a hopeful outlook or desire for the government to take action, indicating a positive inclination towards addressing challenges [+Judgement: Capacity and Propriety]—Evaluates the government's ability and appropriateness in taking the lead by following important reports and recommendations, suggesting competence and responsibility [+Appreciation: Valuation]—Expresses the importance and significance of transitioning to a sustainable economy, appreciating the positive impact this change could have

No	Positive sentiment lexicon (PSL)	Context	Affect (emotions felt)	Judgement (of people's behaviour)	Appreciation (of things)	Description
3	Like	Therefore, the Association for Community and Dialogue would like to appeal to all politicians and civil society leaders [+affect: inclination] to embrace sustainable developmentgoals as the supreme principle [+judgement: proriety] in which all economic and social engineering activities are based on [+appreciation: valuation] as we discuss the 12th MalaysiaPlan from 2021 to 2025. [+judgement: capacity]	Inclination	Propriety Capacity	Valuation	[+Affect: Inclination]—Reflects a desire or hope that politicians and civil society leaders will take action, indicating an inclination towards encouraging certain behaviour [+Judgement: Propriety]—Evaluates the ethical responsibility or appropriateness of adopting sustainable development goals as a guiding principle, suggesting a moral obligation [+Judgement: Capacity]—Implies a call to action, assessing the capability of the stakeholders to implement and discuss these goals effectively within the given timeframe [+Appreciation: Valuation]—Suggests the appreciation of the importance and value of sustainable development goals as the foundation for economic and social activities, emphasising their central role

efficacy of existing solutions, and recognises the significance of technological and creative innovations. Likewise, "opportunity" conveys excitement for governmental initiatives, while also implying accountability and the possible effects of sustainable behaviours. The term "like" signifies a favourable disposition or aspiration for cooperation between politicians and civil society, while simultaneously underscoring the ethical obligation to prioritise sustainable development objectives.

This approach underscores the evaluative capacity of language and demonstrates how these PSLs influence attitudes and behaviours in discourse. The table shows the linguistic framing of attitudes, behaviours, and values, highlighting the function of language in cultivating optimism, responsibility, and appreciation in dialogues concerning urgent societal and global issues.

Appraisal Categorisation: Negative Sentiments and Excerpts in New Straits Times

Table 5.5 depicts the appraisal categorisation of negative attitudes in the New Straits Times, demonstrating how negative evaluations are dispersed across the affect, judgement, and appreciation subsystems. This classification sheds light on how the newspaper delivers critical viewpoints, notably in terms of emotional responses, appraisals of individuals' competency and morality, and assessments of policies and institutional structures.

The most common sentiment in the affect subsystem, which captures emotional reactions, is insecurity, which appears 14 times, indicating that concerns about instability, uncertainty, and potential dangers are widespread in the newspaper's discourse. Dissatisfaction appears 11 times, expressing displeasure with policies, government, or societal issues. Unhappiness is recorded twice, but disinclination appears just once, indicating that direct emotional responses and statements of desire or reluctance are less prominent in the negative sentiment portrayal.

The judgement subsystem, which assesses people's competence, perseverance, and ethical behaviour, is the most common category in the table. The most prominent sub-branch of social esteem is capacity (16

Table 5.5 Negative sentiments categorisation in new straits times

Subsystem	Sub-branches	Level 1 negative frequency	Level 2 negative frequency	Total negative frequency
Affect	(a) Happiness/Unhappiness	2	28	28
	(b) Security/Insecurity	14		
	(c) Satisfaction/Dissatisfaction	11		
	(d) Inclination/Disinclination	1		
Judgement	*Social esteem* (a) Normality (how special/ fate/ usuality)		18	51
	Social esteem (b) Capacity (how capable/ability)	16		
	Social esteem (c) Tenacity (how resolute/resolve/ inclination)	2		
	Social sanction (d) Veracity (how honest, truth/ probability)	26	33	
	Social sanction (e) Propriety	7		
Appreciation	(a) Reaction (impacts and qualities)		26	26
	(b) Composition (balance and complexity)			
	(c) Valuations	26		
Total of appraisal occurrences				105

occurrences), which indicates criticism of the efficiency of persons or institutions. Tenacity (2 occurrences) indicates that a lack of resilience and dedication is a small issue in the debate. Normality does not appear, showing that negative assessments are based on efficacy and competency rather than the expectedness of an event.

The top reported category in the table for social sanction, which examines ethical and moral judgement, is veracity (26 occurrences), showing that concerns about honesty, trustworthiness, and truthfulness are fundamental to the newspaper's unfavourable attitude representation. Propriety (7 occurrences) emphasises ethical considerations and moral responsibilities in governance or leadership.

The appreciation subsystem, which assesses the worth, impact, and complexity of events or policies, has valuation (26 occurrences) as its most dominant sub-branch, indicating that negative sentiment is strongly directed towards assessing the significance and consequences of policies or systemic structures. However, reaction (impacts and characteristics) and composition (balance and complexity) do not show, implying that negative emotion is shaped more by the magnitude of concerns than by their immediate consequences or structural composition.

In summary, judgement has the most negative occurrences (51), showing that New Straits Times focuses mostly on critiquing competence, ethical behaviour, and moral integrity. The affect subsystem follows with 28 instances, emphasising worries about emotional discontent and insecurity. The appreciation subsystem records 26 occurrences, indicating an emphasis on assessing the importance of systemic or policy-related concerns.

This analysis reveals that the New Straits Times generates negative sentiment mostly through criticisms of credibility, governance, and systemic flaws, rather than solely emotional responses. The table summarises the distribution of negative assessments, with more context and analysis provided in the following section. The excerpts' examples derived from New Straits Times are illustrated below.

Table 5.6 critically examines how negative sentiments are articulated in the New Straits Times coverage of environmental issues. Based on Martin and White (2005) appraisal framework, the examples are grouped into the affect, judgement, and appreciation domains to show how emotions, ethics, and values are reflected in the newspaper discourse. Its table shows that linguistic choices place environmental degradation, political inertia, and corporate misconduct not as descriptive facts, but as urgent problems to be condemned and actioned upon.

First example, the lexicon "disaster" is used for "some of the most shocking corporate scandals" involving climate issues. It is one of the negative Affect expressions—indicating dissatisfaction, disappointment, and concern. Meanwhile, the phrase suggests negative judgement, particularly of propriety, casting corporate actors as having failed in their ethical obligations. Extending the discussion to "climate disaster in the history of mankind" adds a negative appreciation to the discourse and the historical gravity and destructive impact of such misconduct. In doing so, the text condemns both the corporate actors and the broader systemic consequences of unchecked industrial behaviour on the climate.

The second example, "lack" is critiqued as "lack of political will". That phrase has a lot of negative Affect, which indicates dissatisfaction and frustration that political leaders are hesitant to act decisively on environmental law. It is further classified as a negative Judgment of tenacity, showing deficiencies in resolve and commitment. Here, political will is reframed not as an administrative defect, but as a moral and motivational failure, when delays are based on economic interests. The associated mention of potential impact on livelihoods, lands, properties, and profits—a negative appreciation—amplifies the critique of ecological integrity and socio-economic stability. This framing makes political hesitation both ineffective and harmful, and supports a narrative of institutional neglect.

The third example is lexical degradation, which describes forest degradation and land clearance dangers. The expression suggests negative affect in the form of insecurity, pointing out that such practices place Malaysia at a greater risk. The tone matches public anxiety about repeated environmental crises like haze pollution. It also triggers negative judgements of propriety, declaring such practices as irresponsible and ethically questionable. The discourse also mobilises negative appreciation through terms such as "onslaught from transboundary haze", assessing the consequences for health, ecology, and national well-being. Such stark

Table 5.6 Examples of new straits times' negative excerpts

No	Negative Sentiment Lexicon (NSL)	Context	Affect (emotions felt)	Judgement (of people's behaviour)	Appreciation (of things)	Description
1	Disaster	On the flipside, last year also highlights some of the most shocking corporate scandals [-affect: dissatisfaction] [-judgement: propriety] relating to climate disaster in the history of mankind [-appreciation: valuation]	Dissatisfaction	Propriety	Valuation	[-Affect: Dissatisfaction]: The phrase "most shocking corporate scandals" conveys a strong feeling of displeasure or disappointment towards the corporate actions linked to climate disasters [-Judgement: Propriety]: "Corporate scandals" suggest morally wrong behaviour, indicating a negative judgement on their ethics and actions [-Appreciation: Valuation]: The phrase "climate disaster in the history of mankind" emphasises the severe and detrimental impact, reflecting a negative evaluation of the events' significance

No	Negative Sentiment Lexicon (NSL)	Context	Affect (emotions felt)	Judgement (of people's behaviour)	Appreciation (of things)	Description
2	Lack	But there has been a lack of political will [-affect: dissatisfaction] [-judgement: tenacity] to implement these laws due to the potential impact on livelihoods, lands, properties, and profits [-appreciation: valuation]	Dissatisfaction	Tenacity	Valuation	[-Affect: Dissatisfaction]: The phrase "lack of political will" conveys frustration or displeasure with the political leadership's failure to implement the laws effectively [-Judgement: Tenacity]: The sentence criticises the lack of resolve or commitment from political leaders, reflecting negatively on their willingness to pursue actions that might be challenging or unpopular [-Appreciation: Valuation]: The reference to "potential impact on livelihoods, lands, properties, and profits" implies a negative outcome, signalling that the failure to implement laws is perceived as having adverse economic and social consequences

(continued)

Table 5.6 (continued)

No	Negative Sentiment Lexicon (NSL)	Context	Affect (emotions felt)	Judgement (of people's behaviour)	Appreciation (of things)	Description
3	Degradation	Forest degradation and land clearing [-judgement: propriety] expose Malaysiato further risks [-affect: security] of home-grown haze in the future alongside the ongoing onslaught from transboundary haze [-appreciation: valuation]	Insecurity	Propriety	Valuation	[-Affect: Insecurity]: The terms "expose Malaysia to further risks" suggest a heightened sense of vulnerability regarding environmental issues, inducing a feeling of insecurity about Malaysia's future air quality [-Judgement: Propriety]: The phrase "forest degradation and land clearing" conveys a judgement about the moral or ethical implications of such activities, hinting at irresponsibility or lack of ethical consideration [-Appreciation: Valuation]: The reference to "ongoing onslaught from transboundary haze" signals a negative evaluation of Malaysia's environmental condition, indicating significant challenges to air quality and public health

language choices render Malaysia as an additional victim and a cause of ecological decline.

In summary, the table shows how the New Straits Times builds negative sentiment via dissatisfaction, ethical judgement, and adverse valuation. The discourse condemns corporations, political leaders, and destructive industries as moral and social failures with serious consequences. The analysis shows how negative language is deployed—not as passive descriptions but as calls to accountability and reform. In this way, the newspaper's reporting forms part of an environmental justice discourse that mobilises emotional, ethical, and evaluative resources for public concern and change.

Appraisal Categorisation: Positive Sentiments and Excerpts in New Straits Times

Table 5.7 shows how appraisals classify positive sentiments in the New Straits Times, showing how favourable evaluations are distributed across the subsystems of affect, judgement, and appreciation. This classification reveals how the newspaper tries to project optimism—through emotional responses, competence assessments, or recognition of the value of policies and systemic initiatives.

On average, in the affect subsystem, positive emotional responses occur most frequently in the category, for instance, security (11 times) and inclination (11 times). They also suggest that stability, safety, and willingness to act are frequently cited in the paper as key features of positive framing. Satisfaction appears five times, suggesting moderate approval of governance or policy outcomes, but happiness only appears once—suggesting that direct expressions of joy or elation are rarely noted.

The judgement subsystem evaluates behaviour and performance against social norms. With 27 occurrences, capacity dominates with high positive evaluations of competence and institutional capacity to act effectively on environmental issues. Tenacity appears only once, and suggests that resilience or perseverance is rarely mentioned. This excludes normality and suggests that positive judgements are more concerned with

Table 5.7 Positive sentiments categorisation in new straits times

Subsystem	Sub-branches	Level 1 positive frequency	Level 2 positive frequency	Total positive frequency
Affect	(a) Happiness/Unhappiness	1	28	28
	(b) Security/Insecurity	11		
	(c) Satisfaction/Dissatisfaction	5		
	(d) Inclination/Disinclination	11		
Judgement	*Social esteem* (a) Normality (how special/fate/ usuality)		28	28
	Social esteem (b) Capacity (how capable/ability)	27		
	Social esteem (c) Tenacity (how resolute/ resolve/ inclination)	1		
	Social sanction (d) Veracity (how honest, truth/ probability)			
	Social sanction (e) Propriety			
Appreciation	(a) Reaction (impacts and qualities)		29	29
	(b) Composition (balance & complexity)			
	(c) Valuations	29		
Total of appraisal occurrences				85

the expectedness of events than with their efficacy and results. Furthermore, also absent are veracity and propriety, which indicate that moral and ethical commendations are not central to the newspaper's positive evaluations.

Appreciation is the subsystem that most influences positive sentiment. The 29 instances of valuation show that praise is most often directed at the perceived worth and significance of policies, systemic improvements, or climate change initiatives. By contrast, reaction (impact and qualities) and composition (balance and complexity) are absent, suggesting that positive discourse is articulated less through immediate effects or structural balance than through broader recognition of value.

In summary, appreciation has the most positive instances (29), followed by Affect (28) and Judgement (28). This relatively even distribution argues that the New Straits Times frames positive sentiment as a balanced mix of emotional engagement, competence evaluation and value recognition, with a slight emphasis on appreciation of systemic significance.

This analysis reveals that the New Straits Times constructs its positive discourse not through overt emotionality but through a complex web of stability, competence, and valuation, highlighting how important impactful policies and initiatives are in the environmental domain.

How positive sentiments are constructed in articles in New Straits Times dealing with environmental themes is assessed in Table 5.8. This analysis reveals how positivity is used to describe environmental efforts as well as to mobilise support, increase confidence in governance, and highlight sustainability campaigns value. The selected examples, using the lexical items "opportunity", "vital" and "better", illustrate how emotional encouragement, competence, and positive valuation are infused into messages of environmental stewardship.

The first example is the word "opportunity", which describes Earth Day as "a golden opportunity to promote green buildings and GHG emissions". This is coded positive affect under inclined for optimism and forward-looking attitude. This is anticipation of Earth Day being the catalyst for environmental change. The sentence also reflects confidence in stakeholders to act, which is classed as judgement in capacity. It casts actors of the green building initiative as competent and effective,

Table 5.8 Examples of new straits times positive excerpts

No	Positive Sentiment Lexicon (PSL)	Context	Affect (emotions felt)	Judgement (of people's behaviour)	Appreciation (of things)	Description
1	Opportunity	The upcoming Earth Day on April 22, therefore, will be a golden opportunity [+affect: inclination] to promote green buildings and reduce greenhouse gas (GHG) emissions [+judgement: capacity] [+appreciation: valuation]	Inclination	Capacity	Valuation	[+Affect: Inclination]: Highlights optimism and eagerness to leverage Earth Day for positive environmental action [+Judgement: Capacity]: Reflects confidence in the capability of stakeholders to promote green buildings and tackle greenhouse emissions [+Appreciation: Valuation]: Recognises the significance of Earth Day as a valuable platform for environmental advocacy

No	Positive Sentiment Lexicon (PSL)	Context	Affect (emotions felt)	Judgement (of people's behaviour)	Appreciation (of things)	Description
2	Vital	Therefore, it is vital for Malaysiato establish an energy-price stabilisation [+affect: security] fund to help consumers [+judgement: capacity] absorb sudden shocks in energy prices [+appreciation: valuation]	Security	Capacity	Valuation	[+Affect: Security]: Suggests increased stability for consumers against unpredictable energy price shocks [+Judgement: Capacity]: Reflects the government's ability to create mechanisms to manage energy price volatility [Appreciation: Valuation]: Highlights the importance of a stabilisation fund in ensuring economic resilience
3	Better	Let's celebrate this year's theme for World Wetlands Day [+affect: inclination] by ensuring a better and sustainablefuture for the world [+judgement: capacity], especially our future generation [+appreciation: valuation]	Inclination	Capacity	Valuation	[+Affect: Inclination]: Encourages proactive participation in celebrating World Wetlands Day [+Judgement: Capacity]: Acknowledges the ability of individuals and communities to work towards sustainability [+Appreciation: Valuation]: Highlights the significance of wetlands conservation for future generations

thereby enhancing public trust in their competence. The description of Earth Day as a "golden opportunity" also assigns high value to the event, which is covered under Appreciation as valuation. The example positions the occasion as symbolic as well as a pragmatic moment for promoting sustainable development. With this combination of emotional appeal/ trust in capacity/event valuation, New Straits Times creates a hopeful/ actionable tone for environmental participation.

A second example uses the word "vital" when proposing an energy-price stabilisation fund. The sentence describes the fund as essential to help consumers weather sudden energy price shocks. It suggests some level of reassurance or protection and is therefore classified as Affect in the security subdomain, reflecting emotional comfort with the future stability. In addition, under Judgement capacity is recognised for the government to create such a stabilisation mechanism, which suggests institutional competence in economic and environmental governance. Finally, the stabilisation fund concept is appraised positively in terms of value, notably for its contribution to economic resilience, and is classified under Appreciation as valuation. This example shows how trust in public institutions and their role in protecting vulnerable communities can be reinforced through positive sentiment while also recognising wider social benefits of sustainable policy measures. The language legitimises state intervention but also reasserts readers of its long-term value.

The third example uses the word "better" in reference to World Wetlands Day. The phrase "ensuring a better and sustainable future" indicates positive Affect as inclination and thus a hopeful desire to take action on environmental issues. Call to action creates participation and collective responsibility. The sentence also expresses confidence that people can contribute to this end, which is assessed under Judgement, in terms of capacity. It recognises that people and communities can create sustainable change. With "especially our future generation" the conservation goal is framed as intrinsically worthwhile and forward looking, so the statement is classified under Appreciation. The wetlands are called ecologically essential but also part of a legacy that must be preserved for those who come after us. In this example, personal responsibility meets intergenerational value, a duty coupled with optimism for long-term environmental impact.

The three examples from Table 5.8 show how New Straits Times constructs a discourse of empowerment, moral obligation, and sustainable progress with positive evaluative terminology. With expressions of emotional optimism, institutional competence, and social value, the newspaper joins its environmental reporting with a broader public discourse that claims hope and agency in the face of ecological challenges. Such discursive choices mobilise, validate, and encourage, not just inform. This Appraisal framework reveals here the affective and ideological dimensions of positive climate communication and how language might be positioned to increase environmental awareness and promote collaborative action. As such, these examples support the theoretical foundations of discourse analysis as well as a lens through which to understand how mainstream media in Malaysia construct a solution-oriented environmental narrative.

Tabloid Versus Mainstream: Underlying Attitudes

The appraisal analyses of Malaysiakini and the New Straits Times (NST) reveal two distinct but complementary evaluative profiles in Malaysian climate change opinion discourse. Both outlets treat climate change as serious and consequential, yet they construct stances through different emphases within Appraisal Theory's subsystems of Affect, Judgement, and Appreciation. In Malaysiakini, negativity is driven first by Affect—especially dissatisfaction and insecurity—producing a discourse of frustration and unease that often foregrounds shortcomings in climate governance and uncertainty over environmental futures. This emotional pressure is closely allied with Judgement, where social sanction via propriety is prominent, and actors are criticised less for incompetence than for ethical failings and irresponsible choices. By contrast, NST's negativity is anchored in Judgement, particularly social esteem (capacity) and social sanction (veracity), which positions climate politics as a problem of credibility and performance. The question is not only whether actors should act, but whether they can and whether they are honest about doing so. In

Appreciation, Malaysiakini tends to highlight immediate impacts (reaction), whereas NST scales up significance (valuation) alongside impact, translating discrete events into systemic stakes for policy, livelihoods, and national well-being.

These orientations come into sharper focus when we examine the lexicons that carry negative evaluations. For instance, Malaysiakini regularly deploys items such as long, serious, critical, disaster, and lack to narrate enduring crisis, ethical breach, and concrete harm, as seen in its treatment of the Sungai Kim Kim incident, where unhappiness (Affect), impropriety (Judgement), and severe consequences (Appreciation–reaction) converge. NST, while sharing core crisis terms like disaster and critical, usually pairs them with lexicon that links environmental decline to social and economic vulnerability; degradation, poverty, concern, and even blame. This pairing sustains a narrative in which environmental damage is simultaneously a governance failure and a societal risk, warranting scrutiny of both competence (capacity) and truthfulness (veracity).

Positivity, though less frequent overall, is constructed differently as well. In Malaysiakini, positive Affect concentrates in inclination, which is a forward-looking desire to act, while Judgement tends to recognise capacity and tenacity, and Appreciation foregrounds the valuation of initiatives that merit endorsement. Positive lexicons such as support, better, hope, opportunity, and advocacy cluster around civil mobilisation, innovation, and the worth of sustainable transitions. NST, meanwhile, distributes positive stance more evenly across subsystems but gives pride of place to Judgement of capacity and Appreciation via valuation, signalling faith in institutional efficacy and the significance of policy pathways. Its positive lexicons; opportunity, vital, better, and a broader set elsewhere (like, great, best, well-being, exemplary)—legitimise action by projecting stability (affect–security), competence (judgement–capacity), and systemic importance (appreciation–valuation).

Taken together, the two outlets articulate different routes to urgency. Malaysiakini arrives there by moralising failure and centring ethical responsibility; its discourse presses from the outside in, amplifying grievance and insisting on accountable conduct. NST approaches by institutionalising competence and scaling up the value of decisions; its discourse reassures from the inside out, presenting pathways through which policy can matter and publics can be protected. Neither stance is simply "negative" or "positive"; rather, each organises negativity and positivity in ways that align with a distinctive communicative role—alternative critique versus mainstream reassurance—and together they constitute a communicative ecology in which pressure and permission, indignation and legitimacy, co-produce a national conversation about climate action.

These contrasts hold implications for how readers are positioned. Malaysiakini's emphasis on dissatisfaction and propriety invites audiences to evaluate climate politics in moral terms and to see themselves as agents of pressure—citizens whose endorsement is contingent upon ethical rectification. NST's emphasis on capacity, veracity, and valuation invites audiences to evaluate climate politics in terms of deliverability and significance—constituents whose support is tied to credible performance and demonstrable societal benefit. In both cases, Affect is not ornamental; it is strategically linked to Judgement and Appreciation to transform facts into grounds for either condemnation or consent.

Finally, taken across the dataset, the two outlets are not mutually exclusive but complementary. Malaysiakini keeps the problem vivid and ethically salient; NST keeps solutions visible and institutionally plausible. The first risks despair unless it makes room for hope and capacity; the second risks complacency unless it keeps scrutiny and accountability in view. Where they overlap, in shared crisis lexicon and occasional convergence on capacity and valuation, we observe the contours of a robust public sphere: one that both demands better and shows how "better" might be achieved.

Conclusion

The appraisal analysis of Malaysiakini shows that its climate change opinion coverage leans heavily on negative affect, particularly dissatisfaction and insecurity, combined with strong judgement of propriety that frames inaction and poor governance as ethical failures. Positive discourse, while less frequent, highlights inclination, capacity, and valuation, signalling aspirations for change, confidence in competence, and recognition of the value of sustainable initiatives. This pattern positions climate change as both a moral responsibility and a site of public frustration, while still allowing space for hope and advocacy.

In contrast, the New Straits Times portrays judgement, especially capacity and veracity, with negative evaluations aimed at questioning institutional effectiveness and credibility. Positive sentiment is more evenly distributed across affect, judgement, and appreciation, with particular emphasis on competence and policy valuation. This reflects a framing that prioritises institutional responses, stability, and the systemic significance of climate initiatives. Collectively, the analysis findings suggest that Malaysiakini stresses ethical urgency, while NST underscores institutional competence, together shaping complementary orientations to climate change in Malaysian opinion discourse.

To recapitulate, these findings show that appraisal resources are not merely linguistic embellishments but central mechanisms through which newspapers shape public perception of climate change. By highlighting different dimensions, Malaysiakini through ethical urgency and public dissatisfaction, and the New Straits Times through competence, credibility, and systemic value, both newspapers reveal how opinion discourse positions readers to feel, judge, and evaluate environmental issues. This underlines the crucial role of evaluative language in constructing climate change as both a moral and institutional challenge, reinforcing the significance of appraisal analysis in understanding media's contribution to climate communication.

Reference

Martin, J. R., & White, P. R. R. (2005). *The language of evaluation: Appraisal in English*. Palgrave Macmillan.

6
Concluding Reflections

Introduction

This final chapter synthesises insights from the preceding five chapters and considers what these might mean for understanding language, climate, and public discourse in Malaysia. While all previous chapters dealt with one aspect of analysis, from corpus-assisted keyword and frequency analysis, sentiment analysis, ecolinguistics analysis, critical discourse analysis, and appraisal analysis, this concluding discussion synthesises these findings into one narrative. The aim here is not to reproduce results, but to draw connections, identify common threads, and place them within larger debates in climate communication and discourse studies.

It is a four-part chapter. The first section summarises key findings of chapter one to five while revisiting methodological insights and empirical trends that shaped this study. In the second section, thematic reflections identify discursive patterns of urgency, optimism, ethics, and ecological worldviews and compare how they are embodied in Malaysiakini and the New Straits Times—the two main samples of newspapers that we used in this study. The third section discusses their impact on policymakers, journalists, educators, and scholars and suggests future directions for

N. F. M. Nor et al., *A Critical Analysis of Climate Change Opinion News in Malaysian Media*, https://doi.org/10.1007/978-3-032-10333-8_6

research. The concluding section offers broader reflections on language as an evaluative tool, tying the book together and reminding us all that discourse shapes climate communication. In this way, the chapter ends the book and points outward with lessons for future inquiry and practice.

Revisiting the Analytical Journey

As mentioned, this concluding chapter synthesises the analyses of the preceding five chapters to reflect on language, climate, and public understanding. The idea behind this book is that language is a mirror of climate reality as much as a constitutive force determining how climate change is seen, valued, and acted on. The Malaysian case demonstrates that opinion news lends itself well to this investigation: it offers evaluative commentary coupled with persuasive positioning, reflecting both journalistic practice and wider socio-political contexts.

Chapter one established that Malaysian media discourse on climate change must be read within a broader political economy of communication embedded in regulatory, ownership, and ideological structures (George, 2012). As a mainstream outlet historically geared towards government and corporate priorities, the New Straits Times frames climate issues in terms of development and governance. Instead, Malaysiakini presents itself as an alternative, independent platform and has a reputation for adversarial or critical reporting (Steele, 2009). This structural divergence is more than background detail; it shapes the discursive strategies both outlets employ to frame climate change.

Chapter two explained why this multidimensional methodological approach proved necessary. Climate discourse is layered and cannot be captured adequately through one lens. Corpus linguistics revealed large-scale patterns of keywords and collocations (Baker, 2006; McEnery & Hardie, 2012). Sentiment analysis maps tonal distributions (Liu, 2020). Appraisal Theory added depth by uncovering the attitude subsystem of Affect, Judgement, and Appreciation (Martin & White, 2005). Critical discourse analysis linked these evaluations to power and ideology questions (Fairclough, 2013). Ecolinguistics examined worldviews underpinning language, asking whether stories reinforce ecologically sustaining values and/or destructive anthropocentric assumptions (Stibbe, 2021). Such methodological pluralism allowed a rich account of the discursive construction of climate change in Malaysian opinion news.

Chapter three, focusing on lexical and keyword strategies, had different emphasis in both newspapers. Malaysiakini adopted critical and activist language such as "disaster", "critical", and "lack", and hopeful terms such as "support". NST used more institutional, technocratic and forward-looking terms like "policy", "government", "future", and "opportunity". They reflected divergent ideologies: Malaysiakini emphasised accountability and critique, NST governance capacity, and systemic solutions.

Chapter four covered sentiment, CDA and ecolinguistics in addition to keywords. Both outlets carried negative sentiment, though negativity performed different functions. The negativity of Malaysiakini reflected dissatisfaction and moral critique about governance failures, corruption, and systemic risks. NST's criticisms were more about institutional performance and credibility on whether government bodies and industries are competent or trustworthy. Importantly, positive sentiment, though less frequent, showed how optimism was shaped. Malaysiakini invoked hope and advocacy to mobilise civil society and grassroots agency, whereas NST defined optimism as systemic stability and confidence in long-term policy commitments. CDA noted that NST reprinted dominant discourses of development, whereas Malaysiakini allowed more oppositional framing. Ecolinguistics revealed that Malaysiakini sometimes used ecocentric stories about care and interdependence, whereas NST was still anthropocentric about human-centred development.

The analysis of appraisal was added in Chapter 5. Malaysiakini's negative evaluations were centred around Affect (mainly dissatisfaction/insecurity) & Judgement (primarily propriety). This illustrated how climate change was seen as a site of frustration and moral accountability. Their positive appraisals grounded inclination, capacity, and valuation, indicating that optimism took the form of aspiration and recognition of importance. NST instead relied heavily on Judgement in its negative appraisals, capacity and veracity dominated, criticisms of institutional competence and honesty. Its positive appraisals were more evenly distributed, with capacity and valuation showing that institutional trust and systemic worth were central to its optimism. These findings rounded out the picture: Malaysiakini conceptualised climate discourse as a moral and political crisis for accountability, while NST constructed it as a challenge of institutional performance and systemic resilience.

In general, all the analyses demonstrate that language in Malaysian opinion news is evaluatively rich, ideologically contested, and ecologically consequential. The contrast between Malaysiakini and NST reveals how alternative and mainstream outlets perform distinct but overlapping roles in climate communication: one as watchdog and moral critic, the other as institutional stabiliser and policy translator.

Themes Emerging from Climate Discourse

A few interconnected themes emerge when we compare the findings of the preceding chapters. These themes explain the linguistic nature of Malaysian climate discourse as well as the ideological stakes in how the public are positioned to think and feel about climate change. They include the urgency discourse, the negativity/optimism balance, the ethics of responsibility, and the ecological orientations embodied in language.

A major theme found is urgency. Climate change is frequently made urgent, though how urgency is constructed differs significantly between Malaysiakini and the New Straits Times. Malaysiakini mobilises urgency via affective/ethical registers. Such words as "disaster" and "critical" place climate change as a catastrophe, and evaluative choices represent dissatisfaction, insecurity, and moral accountability. This corresponds to what Carvalho has argued (2007); media create risk imaginaries that describe climate impacts but also create a sense of crisis. In other words, Malaysiakini makes climate urgency look like an environmental risk and a moral failure. NST builds urgency more institutionally, in contrast. Its discourse places emphasis on governance, systemic stability, and policy action, often making urgency a matter of state responsibility instead of collective outrage. Based on Schäfer and Painter (2021), mainstream media normalise urgency by placing it within governance discourses rather than activist ones.

Another theme is optimism, which tempers the otherwise dominant negativity of climate coverage. Both newspapers mobilise positive sentiment in different ways. Malaysiakini focuses on words like "hope", "support", and "opportunity", which often appear in contexts of civil society action or grassroots advocacy. Such positive evaluations highlight inclination, capacity, and valuation and emphasise optimism as something citizens and communities must enact. In contrast, NST is positive about institutional competence and systemic initiatives. "Opportunity", "better", and "vital" tend to be used in contexts applauding government policies or international events like Earth Day or COP26. In this sense, optimism is defined as emanating from institutions rather than citizens. This contrasts with what Boykoff (2011) terms "discursive balance"; mainstream outlets temper criticism of climate inaction with assertions of systemic stability while alternative outlets play optimism as a means of mobilising counter-hegemonic energies.

These also highlight ethical dimensions of climate discourse. Malaysiakini's habit of making propriety a type of negative judgement demonstrates its tendency to treat climate change as a moral and ethical issue. Corporate misbehaviour, government inaction, and environmental negligence are repeatedly called wrong/ irresponsible/ unethical. This emphasis corresponds to Fairclough (2013)'s idea of discourse as moral positioning, in which evaluative language defines what is acceptable conduct. Rather than relying on veracity in its negative judgements, NST stresses veracity. Questions of honesty, transparency, and credibility dominate, suggesting trustworthiness is at the core of how the mainstream paper frames climate discourse. That also echoes Painter's (2013) finding that credibility and legitimacy are often overemphasised in mainstream outlets when covering science and policy. The two newspapers imply two differing but complementary ethical orientations; One was morality and fairness, the other was truth and credibility.

Another theme concerns ecological worldviews, best understood through ecolinguistics analysis. Malaysiakini sometimes uses ecocentric framings that stress interdependence, care, and responsibility towards nonhuman nature. Examples include biodiversity loss and ecological stewardship discussions which often emphasise the intrinsic value of ecosystems. These discursive choices are in agreement with Stibbe (2021) concept of "beneficial stories", which promote ecological consciousness and sustainable values. In contrast, NST remains largely anthropocentric and frames environmental issues as human impacts, development priorities, and governance responsibilities. Forest degradation, for example, is discussed mainly as a risk to livelihoods, health, and economic stability rather than as a moral or ecological harm in itself. This reflects a trend outlined by Cox and Pezzullo (2016) for mainstream media to favour anthropocentric framings linking climate action to human security and development agendas.

This is partly explained by the institutional and ideological positions of the two newspapers. Tabloid outlets like Malaysiakini are more able to adopt activist framings and hold power accountable. Its evaluation strategies, satisfaction, insecurity, and ethical responsibility, reflect its position as a critical voice within Malaysia's media landscape. By contrast, NST is embedded in mainstream political and corporate structures. Its evaluation strategies of competence, veracity, and systemic valuation are in line with its institutional role as legitimiser of governance/policy. This supports George (2012)'s argument that media in Malaysia are more broadly politically-economically aligned, with alternative and mainstream media performing different but related functions in public discourse.

The Malaysian case echoes global trends too. In other contexts, studies have shown that mainstream outlets depoliticise climate discourse by embedding it in institutional routines whereas alternative outlets stress activism and ethical urgency (Boykoff & Yulsman, 2013; See also: Schäfer & Painter, 2021). Simultaneously, Malaysia's postcolonial and developmental context lends these dynamics particular inflections. For example, the continuing emphasis on development/economic growth in NST reflects Malaysia as a fast-developing economy that balances sustainability with industrialisation. Malaysiakini's critical focus on governance failures reflects the country's struggles for corruption, accountability, and political reform. These findings taken together demonstrate how climate discourse is shaped both globally and locally by political economies and cultural contexts.

Real-world examples contextualise these findings. Articles in Malaysiakini framed the 2021 floods in Klang Valley as a governance failure and moral crisis, with articles condemning inadequate preparation and slow response (Malaysiakini, 2021). Meanwhile, NST highlighted government relief measures and policy responses, institutional competence, and recovery efforts. Additionally, coverage of recurring haze episodes related to transboundary burning often reflected Malaysiakini moral critique of corporate irresponsibility, whereas NST covered diplomatic negotiations and regional governance mechanisms. At international forums like COP26, Malaysiakini highlighted activist voices and civil society demands for accountability, whereas NST reported more on government commitments and international partnerships. These examples support the pattern recognition of appraisal analysis results; Malaysiakini concentrates on critique and accountability, whereas NST concentrates on competence and systemic response.

Collectively, these thematic reflections show that climate discourse is plural in Malaysia. It is the product of competing evaluative directions, institutional alignments, and ecological worldviews. Malaysiakini and NST differ in tone & emphasis but both shape public understanding. One brings readers back to the ethical urgency and systemic failures which call for accountability; the other reasserts readers of governance structures/institutional responses. This play between critique and reassurance, of moral outrage versus systemic valuation is the evaluative terrain of Malaysian climate discourse.

Implications and Future Directions

The findings of the analysis in this book have implications for several stakeholders: The public, policymakers, journalists, educators, and scholars of climate communication. They offer directions for future research into how language, ideology, and climate change relate.

The public's understanding of climate change is shaped by mixed feelings of frustration and hope. The inclination towards negative emotions like dissatisfaction and insecurity suggests that the public is not just passively receiving information but is actively engaged in voicing their opinion about the perceived lack of actions from the government. This negative sentiment is likely to fuel public pressure on the government and other institutions to address climate issues more effectively. The public perceived climate change as a moral issue, not a scientific one, because the public lacks exposure to climate change policies which significantly impedes extensive understanding and engagement. The perception of climate change as a moral issue can lead to stronger social activism and greater blame on leaders or governments as being accountable for their decisions. Meanwhile, the presence of a less positive but hopeful discourse indicates that the public is not entirely sceptical. The recognition of positive government initiatives, even if they are discussed less by the public, provides a crucial space for public engagement. It signals to the government that while there is dissatisfaction, there are among the general public, groups of people who are engaged in sustainable and climate resilient project, to find sustainable and competent solutions, such as solving water woes problem in a small village in Pahang, recycling actions and community garden initiatives. This suggests that effective communication from the government highlighting positive actions and demonstrating genuine commitment can increase public awareness, build public trust, and foster a more engaged and hopeful public about climate change matters. This would foster a sense of individual responsibility and empower the public to become active participants rather than passive recipients of policy.

The results inform policymakers about the communicative credibility of climate policy. Most negative evaluations in Malaysiakini and the New Straits Times focused on competence, credibility, and ethical responsibility. Malaysiakini identified governance failures and ethical failures while NST identified institutional honesty and capacity. This means technical climate measures like flood mitigation projects, renewable energy targets, regional haze agreements, etc., should be judged by the public also on their substantive outcomes, whether they are communicated openly and ethically. As Painter (2013) has shown in his comparative study of climate journalism, credibility is at the heart of publics 'decisions about whether they trust climate policies. Communication credibility may be as important as policy design in Malaysia where distrust of institutions has been eroded repeatedly by political scandals.

The 2021 Klang Valley floods illustrate this dynamic. Malaysiakini called the disaster a governance failure, stressing the lack of preparation and coordination. NST covered institutional relief measures and government commitments extensively. In both instances, evaluation focused on credibility, whether the government could be trusted to save citizens from climate disasters. Broader implications are that communicative failures can undermine policy legitimacy, but communicative successes can build it. Thus, policymakers must invest not only in infrastructure and technical solutions but also in open and credible communication strategies that recognise risks, admit shortcomings, and draw out directions forward (Carvalho, 2007). There is currently a gap between mitigation strategies and adaptation efforts, hence, this gap needs to be bridged. A shift is required from a purely government-led approach towards one that actively engages with communities. Policymakers should work directly with local communities and NGOs to create or co-create policies that are practical and reflective of local needs. The collaboration can ensure that mitigation and adaptation reinforce each other, as local behavioural changes and community initiatives can directly support the national climate change mitigation plans, policies, and efforts. Additionally, the government needs to address the dilemma faced by industry players, who as pointed out by the three experts are somewhat isolated from governmental directives. This calls for clear and consistent policy frameworks that offer incentives and support for transitioning to green practices, such as reducing tax on companies that adopt renewable energy, subsidies for waste management technologies, or grants for developing low-carbon business models. Simultaneously, creating platforms for dialogue and collaboration between the government and private sector can help align market dynamics with national climate goals, ensuring that climate action is seen as a business opportunity rather than a burden. Smaller green initiatives that have potentials can be considered for integration into larger, national strategies. By providing funding, resources, or official recognition to these projects, the government can leverage the innovation and commitment already present within society. This approach would not only amplify the impact of these smaller projects but also foster a more inclusive and nationwide movement towards a climate resilient Malaysia.

The findings for journalists demonstrate the power and responsibility of evaluative language. By using words like “disaster”, “lack”, or “opportunity”, journalists create public emotions, judgements, and values. Malaysiakini, by emphasising moral responsibility & dissatisfaction, holds power accountable. Through its emphasis on competence and systemic value, NST reassures and stabilises. Both roles are needed. As Boykoff and Yulsman (2013) argue, climate journalism must balance critique and constructive communication, risk reporting and solution pointing. This balance may be especially tricky in Malaysia given the close links between media ownership, political power, and corporate interests (George, 2012). There are pressures on independent outlets like Malaysiakini to be critical and from mainstream outlets like NST to legitimise official positions. However, both can contribute meaningfully if they develop transparency, balance, and awareness of the ideological weight of their evaluative choices.

Also involved are educators as well as civil society organisations. The evaluative density of climate discourse revealed herein suggests that media literacy requires more than merely teaching citizens to “read facts”. It must also equip them to critically interpret evaluations. How Affect, Judgement, and Appreciation operate in climate discourse may help readers discern between descriptive information and evaluative positioning. Particularly in contexts such as Malaysia where media ecosystems are fragmented and audiences are often confronted with competing frames of the same issue. Critical media literacy may enable citizens to discern when language is being used to call for urgency, to reassure, to assign blame, or to legitimise policy. As Cox and Pezzullo (2016) argue, citizens being able to decode environmental discourse is the first step towards democratic environmental decision-making.

For scholars, the Malaysian case demonstrates the value of mixed-method approaches. The corpus linguistics identified lexical features invisible to human reading; the sentiment analysis quantified tonal orientations; the Appraisal Theory categorised evaluations; the discursive choice was contextualised in terms of power by CDA and ecolinguistics linked language to ecological worldviews. Together, these approaches provided an empirical as well as interpretive account of climate discourse.

That corroborates Baker et al. (2008) argument for methodological triangulation in discourse analysis. All the methods compensate for their limitations by giving a more complete picture. Future climate communication scholarship could build on this model in other contexts and use it for data sources beyond just climate communications, for instance, social media or multilingual corpora.

The analysis in this book also highlights some gaps that future research could fill. For one, multilingual analysis is needed. This study examined opinion news in English but Malaysia has multilingual media including Malay, Chinese, and Tamil outlets. These outlets might frame climate change differently, using different cultural resources and appealing to different publics. Using a comparative, multilingual corpus might reveal how climate discourse engages ethnicity, religion, and cultural values, dimensions central to Malaysian politics and identity. Schäfer & Painter (2021) point out that climate communication is also shaped by local contexts and Malaysia is a good place to discover this diversity.

The other gap concerns social media. Platforms like Facebook, Twitter (X), Instagram, and TikTok are increasingly important mediums for climate communication, especially with younger Malaysians. The discursive strategies may differ from those of traditional news. Hashtags, memes & short-form videos use humour, irony and visual rhetoric as much as evaluative language. Future work might investigate how appraisal resources and sentiment are deployed in such new media contexts and how they intersect with traditional news reporting. As Anderson (2017) points out, digital platforms have made climate communication more participative yet more fragmented. How traditional and new media interact to shape Malaysian climate discourse is an interesting follow up work.

An additional direction is audience reception research. While this book has concentrated on news texts, it is still open to question how readers take in and respond to evaluative framings. Are Malaysians empowered by Malaysiakini's critical tone or alienated? Is NST's emphasis on competence reassuring or complacent? Audience studies might investigate how urban versus rural, younger versus older, politically active versus less politically active groups interpret climate discourse and how these interpretations shape attitudes and behaviours. Such research might link textual analysis to behavioural outcomes to show how discourse translates to public understanding and action (Nisbet, 2009).

Furthermore, more longitudinal research is possible. Climate discourse is dynamic. It changes as political regimes shift, climate impacts become more severe and global negotiations take place. Analysing how different evaluative frames of climate change have evolved in Malaysiakini & NST may reveal some continuity or change dynamics. For instance, how has discourse changed before and after Malaysian participation in COP26? How did the 2015 Paris Agreement influence local frames of climate policy? What has the recurring disaster landscape shaped—haze, floods? Longitudinal analyses would reveal if discursive patterns remain stable or fluid and how they react to broader socio-political and ecological changes.

In summary, the implications are practical and scholarly. Policymakers must recognise the centrality of communicative credibility; journalists must reflect on their evaluative choices; educators must promote critical media literacy; and scholars must continue to innovate methodologically while expanding into new contexts. Further research using these foundations should involve multilingual analysis, social media discourse, audience reception, and longitudinal tracking. When taking all of these into account, these efforts might provide a more nuanced account of how language shapes climate communication in Malaysia.

Final Reflections: Language, Power, and Public Understanding

Bringing together the findings of this book, it becomes clear that language does not merely report on climate change—it frames, evaluates, and shapes how the public come to understand it. In five chapters, we went from lexical strategies to sentiment profiles, appraisal resources to critical and ecolinguistics analyses. Each method yielded complementary insights that taken together offer a portrait of how Malaysian media construct climate change as crisis and opportunity, site of ethical responsibility and institution competence, of fear and of hope.

A comparison of Malaysiakini to the New Straits Times explains this duality. Malaysiakini is a voice of critique that brings out dissatisfaction, insecurity, and ethical propriety. Its discourse focuses on governance failures, corporate misconduct, and moral responsibility and is often affectively driven. That is consistent with its role as a watchdog outlet. In turn, NST offers an instrument for institutional reassurance by focusing on capacity, veracity, and systemic valuation. The discourse is focused on competence, policy action, and systemic stability, in keeping with its role as a mainstream outlet within the political economy of development. These outlets build a dialectic of climate discourse in Malaysia: One is demanding accountability, the other is projecting stability.

However, how should this translate to a public understanding? It first demonstrates that climate communication is highly evaluative in Malaysia, where publics are exposed to descriptions of climate impacts alongside competing evaluations of responsibility, competence, and morality. It also shows that optimism, though less common than negativity, is strategic: Malaysiakini mobilises hope and opportunity for civic action, whereas NST frames optimism as institutional initiatives. Next, it reveals uneven distributions of ecological worldviews: Malaysiakini sometimes uses ecocentric frames of interdependence, whereas NST is anthropocentric. Together, these patterns indicate contested and complementary climate discourse where different outlets serve different communicative functions.

In reflecting on this book, we realise just how discursive choices matter. Words like "disaster", "hope", or "opportunity" encode ideological positions, affect emotional responses and point the public in particular directions regarding climate change. Exploring such choice analyses through corpus linguistics, Appraisal Theory, CDA and ecolinguistics has revealed how evaluative meanings are constructed across scales, from keywords to ideological narratives. This multidimensional approach shows that quantitative rigour can be useful alongside qualitative interpretation: climate discourse is also about numbers, stories and values.

Additionally, the consequences go beyond academia. For policymakers, the findings stress that credibility and transparency are fundamental ingredients for climate governance. For journalists, they call for reflexivity in evaluating choices. For educators, they call for media literacy that enables citizens to decode evaluation. For civil society, they show how language can inspire or resist. Lastly, for citizens themselves, they show climate discourse is something to be shaped, challenged, and reshaped.

Looking forward, climate communication research must innovate methodologically, contextually, and theoretically. This means, methodologically, big-data approaches should be combined with close text analysis. Contextually, it means going beyond Anglophone and Western contexts to multilingual, postcolonial, and Global South perspectives. It means, theoretically, maintaining links between discourse and issues of ideology, ethics, and ecology. The Malaysian case is one contribution to this larger project aiming at showing how local media reflect and shape global debates while embedding them in national political and cultural contexts.

In closing, the central finding is that climate communication is never neutral. Every word, word play, and evaluative choice shapes how the public perceive climate risks, responsibilities, and possibilities. Malaysiakini and New Straits Times both tell different stories. However, both are equally important: One reminds us of accountability and crisis, the other of stability and governance. Together, they show how language, climate, and public understanding interact. We hope this book will contribute both academic knowledge and practical reflection, helping journalists, policymakers, educators, and citizens question how we use words to tell the most urgent story of our time.

References

Anderson, A. (2017). *Media, environment and the network society*. Palgrave Macmillan.

Baker, P. (2006). *Using corpora in discourse analysis*. Continuum.

Baker, P., Gabrielatos, C., KhosraviNik, M., Krzyżanowski, M., McEnery, T., & Wodak, R. (2008). A useful methodological synergy? Combining critical discourse analysis and corpus linguistics to examine discourses of refugees

and asylum seekers in the UK press. *Discourse & Society, 19*(3), 273–306. https://doi.org/10.1177/0957926508088962
Boykoff, M. T. (2011). *Who speaks for the climate? Making sense of media reporting on climate change*. Cambridge University Press.
Boykoff, M. T., & Yulsman, T. (2013). Political economy, media, and climate change: Sinews of modern life. *Wiley Interdisciplinary Reviews: Climate Change, 4*(5), 359–371. https://doi.org/10.1002/wcc.233
Carvalho, A. (2007). Ideological cultures and media discourses on scientific knowledge: Re-reading news on climate change. *Public Understanding of Science, 16*(2), 223–243. https://doi.org/10.1177/0963662506066775
Cox, R., & Pezzullo, P. (2016). *Environmental communication and the public sphere* (4th ed.). Sage.
Fairclough, N. (2013). *Critical discourse analysis: The critical study of language* (2nd ed.). Routledge.
George, C. (2012). *Freedom from the press: Journalism and state power in Singapore*. NUS Press.
Liu, B. (2020). *Sentiment analysis: Mining opinions, sentiments, and emotions*. Cambridge University Press.
Martin, J. R., & White, P. R. R. (2005). *The language of evaluation: Appraisal in English*. Palgrave Macmillan.
McEnery, T., & Hardie, A. (2012). *Corpus linguistics: Method, theory and practice*. Cambridge University Press.
Nisbet, M. C. (2009). Communicating climate change: Why frames matter for public engagement. *Environment: Science and Policy for Sustainable Development, 51*(2), 12–23. https://doi.org/10.3200/ENVT.51.2.12-23
Painter, J. (2013). *Climate change in the media: Reporting risk and uncertainty*. I.B. Tauris.
Schäfer, M. S., & Painter, J. (2021). Climate journalism in a changing media ecosystem: Assessing the production of climate change-related news around the world. *Wires Climate Change, 12*(1), Article e675. https://doi.org/10.1002/wcc.675
Stibbe, A. (2021). *Ecolinguistics: Language, ecology and the stories we live by* (2nd ed.). Routledge.
Steele, J. (2009). Professionalism online: How Malaysiakini challenges authoritarianism. *International Journal of Press/politics, 14*(1), 91–111. https://doi.org/10.1177/1940161208326465

Index

N. F. M. Nor et al., *A Critical Analysis of Climate Change Opinion News in Malaysian Media*, https://doi.org/10.1007/978-3-032-10333-8

FSC
www.fsc.org
MIX
Papier | Fördert gute Waldnutzung
FSC® C083411

Kolibri 360 GmbH
Ferdinand-Jühlke-Straße 7
99095 Erfurt, Deutschland
produktsicherheit@kolibri360.de